Evaluating the Human Environment

Essays in applied geography

Evaluating the Human Environment

Essays in applied geography

John A. Dawson
Lecturer in Geography, Saint David's University College,
University of Wales

John C. Doornkamp
Lecturer in Geography, University of Nottingham

Edward Arnold

© Edward Arnold (Publishers) Ltd. 1973

First published 1973 by
Edward Arnold (Publishers) Ltd.
25 Hill Street, London W1X 8LL

ISBN: 0 7131 5676 7 boards
      0 7131 5677 5 paper

Text set in 11/12 pt. Monotype Baskerville,
printed by photolithography, and bound in Great Britain at
The Pitman Press, Bath

Contents

Acknowledgements

Permission to adapt the following figures is gratefully acknowledged from:

J. B. Caird for figure 2.1; W. Allan for figure 2.3; J. Rutherford, M. I. Logan and G. J. Missen for figure 2.4; R. M. Prothero for figure 2.5; H. Thorpe for figure 2.7; G. T. McDonald for figure 3.5; G. Fisk for figure 6.1; A. Garnett for figure 8.1; W. Bach for figure 8.2; T. J. Chandler for figure 8.3; P. Anderson for figure 8.4; A. Martin and F. R. Barber for figure 8.5.

Professor Douglas gratefully acknowledges comments and suggestions on Chapter 3 by Peter Crabb and Maureen Douglas. The editors wish to record their thanks to Dr. D. A. Davidson for his many helpful comments, to Mr. M. Cutler for drawing the majority of figures, and to Miss R. Williams for typing the text. Grateful acknowledgement is made to Saint David's University College Pantyfedwen Fund for a grant towards the cost of the index.

Introduction

A student of geography inevitably questions the relevance of what he is studying. In particular he is concerned with the practical application of his geographical knowledge and expertise to problems that affect the lives of people. This volume presents eleven studies which illustrate the usefulness and relevance of the geographer's approach to some of the issues which concern man and are related to his environment.

The essays included in this book do not, however, aim to provide exhaustive coverage of the range of application open to geographers. They have been chosen to encompass as broad a spectrum of geographical techniques as is possible within the confines of a short volume, while allowing each topic to be dealt with in some depth. Although the range of approach is wide, three basic and important themes emerge. First, and most important, each chapter shows the relevance of geography to *live* issues. Secondly, they reveal that the many processes of landscape and environmental change can only be understood by consideration of physical and human geography in interdependence. Thirdly, they emphasize the value of the geographer's contribution to interdisciplinary investigations, while also recognizing the importance to geography itself of working on an interdisciplinary basis.

Geographical attitudes are evolving in two major directions, both of which are central to the basic concept of geography as a synthesizing spatial discipline. On the one hand there is an ever-increasing movement towards the *explanation* of geographical processes, whilst on the other there has been, in the last decade, an increased general awareness that the geographer can occupy a major position in the planning and development of human environments. This new awareness, coupled with the equally important change in geographical philosophy from a descriptive art to an analytical science, has produced geographical studies of direct relevance to workers in disciplines whose aim is 'environmental management'. The feedback from such studies, in the form of problems encountered by environmental planners at all administrative levels and working on problems at all scales, has encouraged geographers in their analyses of landscape changes and processes relevant to particular environmental issues. All of this has led to an increased expertise in recognizing and defining trends in an ever-changing environment. This expertise can be extremly useful in building predictive models upon which sensible planning may be based.

In the era since 1945 many governments and development agencies have been anxious that changes in rural areas should not proceed without viable plans. The production of such plans and their sensible application has necessarily to be based on an accurate assessment of the properties of particular areas. Rural land evaluation (Ch. 1) thus leads to a definition of the physical ability of the land to support new development. This may be achieved through the definition of both the land potential and physical limitations. Such evaluation involves identification, mapping and analysis of the elemental properties characterizing both the physical landscape (for example relief, drainage, soils) and the human landscape (for example population distribution and transport network).

Frequently the planned developments in rural areas are dependent on agriculture. Major changes in land usage are an integral and vital part of these rural developments. Sometimes these land-use changes occur in virgin territory; at other times there is planned change of existing land uses. It is pertinent therefore, as Chapter 2 shows, that an accurate statement of existing land use and social structure should be made before bringing about wholesale or even minor—though locally significant—changes in rural areas.

Whether the landscape occupied by man is rural or urban, highly industrialized or of poor agricultural quality, there is a need for an adequate supply of water (Ch. 3). The water resources of the world vary widely, in both quality and quantity, from place to place on a global and a local scale. Water availability depends not only on the physical factors of relief, climate, geology and hydrology, but also on human factors such as the economics of the water supply system, the legalities of water acquisition, and the situation and site of the locations requiring water. Urban area requirements for water are large, and need to be met not only from bore holes but also from reservoirs and storage areas, frequently located in rural areas. Many geographical factors thus influence the water resources available to meet the increasing demand.

Throughout the world, as the range of human activities increases and social and economic patterns become more complex, so problems of the spatial organization of these activities acquire greater significance. Increasingly man works, plays and exists within the urban milieu. The complex interaction of economic and social factors is concentrated within the small area of urban land uses. Organization and control of land use is central to urban culture, and thus the evaluation of land for urban uses is critical to the improvement of the quality of urban life. The analysis of existing urban land-use patterns is an important part of the evaluation procedure necessary if man is to make the best of a portion of the world he has created for himself (Ch. 4).

As populations increase and social patterns become more complex, so there is increased necessity to acquire expertise in the analysis of the order—or disorder—existing in space and the consequent problems. Home and workplace are spatially separated; social planner and industrialist alike have therefore to consider the needs and character of the industrial workforce (Ch. 5). Production and consumption are spatially separated; the resolution of this spatial disparity is a role of the physical planner and marketing specialist alike (Ch. 6). Both labour management and marketing, however, take place against the background of politics and political decision-making (Ch. 7). Through politics, the policies and prejudices of individuals and groups influence the environment which sustains politician, worker and consumer alike.

Man's influence on the landscape may have unexpected and sometimes disastrous side-effects. The pollution of the atmosphere is one such environmental hazard (Ch. 8). As knowledge concerning disease-propagating habitats increases, so does their control and elimination; but man is also creating new environments—city ghettos, broken social structure—which bring about new health hazards additional to those found in the natural enviroment. (Ch. 9).

The speed of change affecting all aspects of society means that the geographer can no longer be content with description and even with static analysis of spatial processes. The spatial processes themselves are changing as the life-style of man changes—as, for example, increased leisure allows increased participation in an ever-widening range of recreational activities (Ch. 10). Recreation itself, however, produces its own environmental problems, not least in the area of conservation (Ch. 11). The development of ideas in the study of conservation has led to a forward-looking approach to the study of man across the globe. Such an approach is fundamental to human survival, and requires the fullest involvement of the student of geography. It accentuates the need for the geographer to be concerned with explanation as well as description, and for the study of process to be central to his studies. The logical development from this study of processes and the explanatory approach is towards a controlled prediction of the future. Thus in geographical work the predictive element, alongside the fundamental analytical one, is being increasingly applied to problems which concern man and his environment.

1 Rural land evaluation

Anthony Young

Land evaluation is the process of estimating the potential of land for one use or several alternative uses. The potential may be given in qualitative terms, as degree of suitability for various forms of land use; in quantitative physical terms, for example as predicted crop yields; or in economic terms, expressed as gross or net cash output.

There is no difference in principle between the evaluation of rural and urban land, but in practice the distinction is usually clear. Thus climate and soil are of relatively low significance for urban land use but are major considerations in rural land evaluation. Two different sets of procedures have been developed, of which that concerned with urban environments is discussed in Chapter 4. In the case of proposals for new urban building construction on previously agricultural land, the decision-making process requires a resolution of the rural and urban land evaluation procedures.

Evaluation is a form of land classification, but there are many types of land classification—for example maps of soils, climate, land

systems—which do not contain any element of evaluation or interpretation in resource terms. Land classifications of the latter type
will not be considered here; a review, with special respect to the
relative merits of integrated and parametric methods in resource
survey, has been given by Mabbutt (1968).

The land use at any given place and time results from decisions
based on the interaction of five groups of factors: environmental,
technological, economic, social and political. The first group comprises the factors of the natural environment: geology, landforms,
climate, hydrology, soils, vegetation and fauna. The second is the
state of technology, for example agricultural or engineering methods,
either at present or at some past time when the decision to place the
land under its present use was made. Economic considerations are of
a different nature, having less physical reality than the two preceding
groups; moreover, economic factors appear as both independent
and dependent variables in the evaluation process. The term social
factors is used in a broad sense to include, besides the relationships of
human groups, considerations of education levels and land tenure
systems. Political considerations arise particularly in the making and
enforcement of planning regulations. Land evaluation is a link which
interprets the physical environment in terms of its resource potential.
This in its turn can be translated into economic terms. By means of
physical planning and productive operations, such as farming or
road construction, these potential resources may then be used.

I Planning environments and purposes

Decisions about land use were made in the past by private individuals.
The evaluation procedure was one of subjective judgement, based on
incomplete information but often incorporating a high degree of
skill born of experience. Where judgement erred, trial-and-error
supervened. It was by such means that the cultivation of particular
crops became localized, long before scientific knowledge of climatic
and soil conditions existed. Moreover, the experienced farmer's
assessment of what will grow where is still frequently better than any
land classification scheme that has yet been devised.

As a result of the great increase in governmental planning powers
arising during and since the Second World War, national and local
planning authorities now have substantial controls over land use
(Ch. 2). These powers are exercised either positively, through government sponsored development planning, or negatively, through legally
enforcible restrictions on land-use changes. It is primarily out of the

need that arises in all forms of administrative control for some uniform basis for decision-making that systematized land-evaluation procedures have been devised.

1 *The need for land evaluation*

Since land evaluation exists mainly for the purpose of planning, a prerequisite for considering evaluation is some discussion of the types of planning decisions required. It is convenient to identify four main world units, which will be called planning environments (see also Ch. 11, II, 3). The distinction between them is based on standards of living and densities of settlement. These are:

(i)   under-developed and relatively sparsely settled lands
(ii) under-developed densely settled lands
(iii) less densely populated lands with a high standard of living
(iv) densely settled developed lands.

The under-developed world may be divided into the sparsely settled continents of Africa and South America, and the more densely settled sub-continents of South and East Asia and Central America. The main land-use planning decisions in the under-developed world are concerned with land development. Low standards of living allow labour-intensive methods of land use to be used. Indeed, there is a school of thought that places maximization of employment opportunities above return on investment as a criterion of development planning. Capital investment, the finance for which comes frequently from overseas aid, is severely limited.

In sparsely settled under-developed regions land settlement of areas previously unused, or used at low intensities, is a major form of development. This planning environment gives rise to the most straightforward type of land evaluation: assessment of the type of land use that will give the maximum return, either per unit area or per unit of capital invested. In the densely settled regions there are only limited opportunities for land settlement, and reorganization of land use on areas already occupied must be the main form of development. Land evaluation here needs to assess the difference in productivity between existing and proposed future uses of the land.

Sparsely settled lands situated in countries with high standards of living, such as large parts of Canada and Australia, give a third context for land-planning decisions. Development can be of the land-settlement type, but is usually privately financed. It involves high capital investment but comparatively small amounts of labour.

Land development must compete for capital with other investment opportunities, and hence profitability is the dominant concern in evaluation.

The last unit comprises the densely settled lands with high living standards, exemplified by Western Europe. In this planning environment a quite different group of planning decisions have come into existence in the past thirty years. Land is in short supply, highly valued, and has a potential for several alternative uses. Planning is primarily concerned with decisions between competing demands for land. In this context it is necessary in land evaluation to devise means of comparing the benefits to be derived from very disparate forms of land use, not all of which can easily be converted into economic terms. Moreover, all land is already used, hence the problem becomes one of re-evaluation for redevelopment.

This division into planning environments is only approximate. Many exceptions occur. For example, there are many densely settled regions of South America and Africa, and a few sparsely settled areas remaining in South Asia, for example in the Dry Zone of Ceylon. South-east Asia contains marked contrasts in settlement densities. Substantial parts of Canada face planning decisions of the type similar to those of Western Europe. Moreover, cutting across these regions are political divisions, between the communist and non-communist worlds, and contrasts between predominant land-tenure systems, as for example between South America, Africa and Asia. Despite its imprecise nature, however, the concept of the planning environment forms a useful context for assessing land-evaluation procedures. The types of decisions to be taken in each planning environment are different, and call for differences of procedure or emphasis in land evaluation.

2   Purposes of land evaluation

Land evaluation may be required for a variety of purposes including:

(a) *Agriculture*
   (i)   Cultivation of a specified crop or crops, annual or perennial
   (ii) Irrigated agriculture
   (iii) Arable farming
   (iv) Grazing, of unimproved or improved pastures
(b) *Forestry*
   (i)   Logging of natural forest
   (ii) Forest plantations

(c) *Water resources*
   (i)   Groundwater and surface water
   (ii) Resources for irrigation
(d) *Mining*
(e) *Engineering purposes*
   (i)   Transportation purposes in general
   (ii) Building foundations
   (iii) Urban use
(f) *Recreation*
(g) *Wildlife conservation (flora and fauna)*
(h) *Special purposes*
   (i)   Military purposes
   (ii) Local government administration (e.g. assessment for taxation)
(i) *Multiple purposes*

Planning in under-developed countries is mainly concerned with the evaluation of land resources. Where mineral reserves exist, their exploitation is normally given absolute priority over agriculture. The main engineering problem in these countries is the construction or improvement of a rural road system.

The competition between urban growth and farming is the major land-use conflict in the densely settled developed lands. The suitability of the land for urban construction (e.g. foundation properties) is not of prime importance; in so far as it is determined at all by considerations of land quality, the planning decision is determined by the agricultural quality. Land evaluation needs to find means of resolving the competing demands of hill farming, forestry, recreation, wildlife conservation and military training.

Among special purposes, evaluation procedures have been developed for military and taxation requirements. The military require forecasts of terrain conditions in territory to which there is no field access. Land evaluation for taxation purposes was early practised, in an elementary form, in India; more recently, it has been the incentive for the development of refined systems in the U.S.A. Other special-purpose evaluations are those requiring a particular combination of conditions on one site (e.g. for the location of dams or airfields).

General and multi-purpose evaluations take into account all, or a range of, forms of land use. The distinction made here is that general-purpose evaluations grade land into a single order, from best to worst, irrespective of specific purposes. Multi-purpose evaluations classify land separately according to its suitability for each of several purposes. General-purpose land classification is of use mainly at the national or larger regional level (Robertson, Jewitt, Forbes and Law, 1968). The apportionment of land between different purposes at the local level requires multi-purpose classification, whilst detailed operations within areas of each land-use type (e.g. farm layout or management of recreational areas) require specialized evaluation procedures adapted to local conditions.

II Approaches to land evaluation

1   *The standard approach*

In the standard approach, land-development planning is regarded as consisting of three phases: description, appraisal and development. The descriptive phase comprises natural resource surveys, for example hydrological investigations, soil survey and forest inventory. The appraisal, or evaluation, phase combines the environmental data with information from technology, such as agricultural methods and crop requirements, and expresses it as resource potential. The development phase is concerned with the physical planning necessary to convert this potential into production, incorporating social and economic factors. These factors are significant, for example, in the evaluation of water resources (Ch. 3).

Most land-evaluation systems are based on the assumption that evaluation occupies a middle position in a sequence of operations, receiving an input of information about the physical environment and providing an output in the form of advice to those concerned with the economics and implementation of development. The three phases are more or less successive in time and, at the end of each, part of the information gathered is passed on in the form of maps; for example, soil, landform, vegetation or land-systems maps at the conclusion of the resource survey phase; land-capability classification maps following evaluation; and development plans at the conclusion of the final phase. The evaluation procedures discussed in later parts of this chapter assume that this approach has been followed. It has, however, been challenged on two counts, and two alternative approaches will first be considered. (This section draws in part on a previous discussion by Farmer, 1971.)

2   *The economic approach*

The first challenge is in effect a suggestion that the sequence environmental survey–technological appraisal–economic evaluation should be reversed. This suggestion was made by Davidson (1965) specifically as a criticism of the methods used by the Division of

Land Research and Regional Survey of C.S.I.R.O. Australia. This organization has surveyed the physical environment of large areas in the northern and interior parts of Australia and in Papua. The results are presented in a series of maps and reports (C.S.I.R.O., 1953–71). It was for these surveys that the land-systems approach was first developed (Christian and Stewart, 1953; Brink, Mabbutt, Webster and Beckett, 1966). The information presented consists largely of descriptive accounts of the physical environment, together with much briefer evaluation sections (exceptions are certain more specific reports on grazing resources). There is no economic analysis.

Davidson argued that most of the lands surveyed could be shown to have little or no potential for development on an economic basis; specifically, investment in land development in the Australian tropics would be unlikely to give as high a return as investment in already settled areas further south. The surveys were therefore collecting information which was of no use as a guide to investment, and so were a waste of government money.

Leaving aside the aspects of this argument that relate specifically to Australian conditions, it is instructive to examine the implications of the alternative approach that Davidson suggested. First, those forms of investment that gave a sufficiently high return to compete with other demands for capital would be determined. This involves assessing demand and market prices for crops and other products, and their costs of production; the latter aspect would involve consideration of technology. Secondly, this analysis would lead to the conclusion that certain types of crops would repay investment, given the present state of technology, and provided that yields were above a certain figure while production costs remained below a corresponding level. The environmental conditions requisite for such levels of output and costs, such as slope and soil properties and drainage or irrigation requirements, could then be determined. Finally, these specific conditions could be sought in the field, with a great saving of field survey costs.

Taken literally, such a method would have disadvantages at least equal to those of the conventional approach. Economists would spend much time demonstrating how profitable it would be to produce numerous crops which there was no possibility of growing (e.g. temperate crops in tropical countries). Natural scientists would find themselves in the position immortalized by Charter (1957) of being 'pedological procurers who are given the task of finding attractive virgin lands for agricultural rape', with no say as to the nature of the resources offered, or how the complex, interacting system of the physical environment might usefully be modified.

This *reductio ad absurdum* of the backwards, or economics-first, approach serves to show up by analogy some deficiencies of the conventional type of survey. Natural resource surveys are prone to collect large amounts of data, much of which is never used in evaluation, still less at the development phase. On the backwards approach, in theory nothing but information relevant to development would be collected. In practice, however, much resource data of relevance to development potential would be missed through the economic blinkers worn.

The resolution of the mutual disadvantages of these two approaches demands greater communication between the various specialists concerned at all stages in the survey. In surveys intended to lead to investment, as distinct from those carried out on a national inventory basis, provisional economic assessments can usefully precede fieldwork; the results could serve as a guide to those areas—and environmental factors—that are likely to require high-density survey. Conversely, the common practice of transferring the natural scientists to another job half way through the project, after they have 'presented their results', is to be deplored; a continued appraisal of the mutual interaction between environmental resources and proposed land use is necessary if best use is to be made both of environmental and economic information.

3   The approach of contemporary functional relationships

The second challenge to the conventional approach was made by Moss (1968, 1969). His alternative proposal is termed that of contemporary functional relationships, or the biocenological approach. He argues that in densely settled tropical areas the distinction between vegetation and land use is not meaningful. In such areas all parts of the land are used by man for productive purposes, whether by cropping or other methods, and this use has caused greater or lesser modifications to the vegetation and soils. Conventional land-use mapping is based on a static model. In reality, every area of land consists of a geo-ecosystem, components of which include the landforms, soil, soil moisture, vegetation and the cropping pattern, including rotational bush fallowing, over a period of years. Each of these components affects the other, through a series of contemporary functional relationships. Any given component may be in a steady state (i.e. unchanging over time), or in a state of change. For example, the soil organic matter content and soil nutrient levels will vary with the ratio between cropping and fallow periods; a perennial crop may either contain proportions of recently replanted, mature and ageing trees such that its overall age remains constant, or there may be insufficient replanting, so that the tree population is ageing with a corresponding reduction in yields. Both these examples involve inter- actions between the economic as well as the ecological systems. The present ecological relationships having been ascertained, these would be used to predict the effects on the environment of proposed changes in land use.

Moss argues his case convincingly with respect to the area of South- west Nigeria on which it was based. He advocates that the approach of contemporary functional relationships should replace the static model of the land-systems method, and that ecology should replace geomorphology as the basis for resource assessment. What is not yet clear is how this approach can be translated into a practical survey procedure, capable of relatively routine application to a variety of areas. It is valuable, however, in emphasizing the fact that the en- vironment land-use relationship is a two-way interaction, and hence that environmental resources are not static, and in drawing attention to the possibility of predicting the effects of land development upon the ecosystem. One respect in which this approach could be applied is to estimate the future organic matter and nutrient status of the soil under defined cultivation systems, for the accomplishment of which a body of information and the outline of a methodology exists (Nye and Greenland, 1960).

In routine surveys neither the economic nor the ecosystem approa- ches are likely to supplant the standard method of survey—appraisal— development. Both the alternatives throw light, however, on the disadvantages of a rigid adherence to compartmentalized surveys with a unidirectional flow of information, and show ways in which the conventional method can be improved by a greater exchange of information between all aspects and at all stages of the survey, from reconnaissance through to the formulation and implementation of development plans.

III Land capability classification

1   *The USDA system*

A convenient basis for examining the normal means of land evaluation, in which appraisal follows resource survey, is the Land Capability Classification used by the Soil Conservation Service of the U.S. Department of Agriculture (Kligebiel and Montgomery, 1961;

hereafter called the USDA system). It illustrates many of the principles involved, and it has been widely applied in developed and underdeveloped countries. Many local and national schemes have been based upon adaptations of it.

The USDA system is an interpretative grouping of soil mapping units, made primarily for agricultural purposes. *Soil* is in fact taken in its wider meaning of *land*, since slope angle, climate and frequency of flooding are taken into account. The main concept used is that of *limitations*, restrictions upon the type of land use or the land potential. A distinction is made between permanent and temporary limitations. Permanent limitations are those which cannot be altered, including slope angle and soil depth; temporary limitations include low soil fertility and minor drainage impedance, each of which can be modified by land management. Land is classified mainly on the basis of permanent limitations. The level of management assumed is that 'within the ability of a majority of the farmers', a mode of definition that enables the application of the scheme to be modified according to farming standards in different countries.

(*a*) *The structure of the USDA system.* On a basis of soil mapping units, there is a three-category structure.

(i) A *capability unit* is a grouping of soil mapping units that have the same potential, limitations and management responses. All soils within a given capability unit can be used for similar crops, require similar management practices and soil conservation measures, and have a comparable productive potential; specifically, the yield range of a crop within a capability unit is not expected to exceed 25 per cent.

(ii) A *capability subclass* is a grouping of capability units that have the *same kinds* of limitation or hazard. These kinds are indicated by letter subscripts, of which in the original system there are four: erosion hazard (e), excess water (w), soil root-zone limitations (s) and climatic limitations (c). Later modifications of the scheme commonly employ additional kinds of limitation (e.g. stoniness, low fertility, salinity).

(iii) A *capability class* is a grouping of capability subclasses that have the *same relative degree* of limitation or hazard. Classes are indicated by roman numerals, the limitation to the type of land use and the risks of damage to the environment increasing from Class I to Class VIII.

The following are abbreviated definitions of the capability classes:

Class I   Soils with few limitations that restrict their use.

Class II   Soils with some limitations that reduce the choice of plants or require moderate conservation practices.

Class III   Soils with severe limitations that reduce the choice of plants or require special conservation practices, or both.

Class IV   Soils with very severe limitations that restrict the choice of plants, require very careful management, or both.

Class V   Soils with little or no erosion hazard but with other limitations impractical to remove that limit their use largely to pasture, range, woodland, or wildlife food and cover. (In practice this class is mainly used for level valley-floor lands that are swampy or subject to frequent flooding.)

Class VI   Soils with very severe limitations that make them generally unsuited to cultivation and limit their use largely to pasture or range, woodland, or wildlife.

Class VII   Soils with very severe limitations that make them unsuited to cultivation and restrict their use largely to grazing, woodland, or wildlife.

Class VIII   Soils and landforms with limitations that preclude their use for commercial plant production and restrict it to recreation, wildlife, water supply or esthetic purposes.

(b) *Relationships within the USDA system.* The main features of this system are its three-category structure; that it is based on negative limitations rather than positive potential; and that the class to which land is allocated is strongly influenced by considerations of soil conservation. At the class level it results in a single ordering of relative value, with a major distinction between cultivable and non-cultivable land between Classes IV and V. It implicitly assumes a decreasing order of value from arable use through grazing and forestry to recreation, wildlife conservation and water-catchment uses, and hence is only valid where these relative priorities hold good. Features of location, such as the distance to markets, are explicitly left out of consideration. No account is taken of the scarcity value of a particular type of land in a given location, a factor which in certain circumstances can greatly modify land values. For example, the headwater catchment area of a river in a region of savanna climate serves to maintain rural water supplies for a large area downsteam; a cliff suited to rock-climbing located near a large city in a lowland area acquires a recreational value out of all proportion to its inherent characteristics.

The meaning of subclass letters changes according to the class to which they are attached. Thus the 'e' in Subclass IIIe indicates a more severe erosion hazard than that in IIe. With the exception of soil depth, the original system does not give precise limits for subclass and class allocation. This lack of precise criteria for the environmental parameters permitted in each class could lead to subjectivity and looseness of definition. However, there is an objection in principle to the use of rigid limiting values, in that the effect of an individual environmental factor varies according to its interaction with others. Thus erosion hazard is not a function of slope angle alone, but of the combined effects of slope angle and length, soil permeability and structural aggregation, and frequency and intensity of rainfall. Even more complex are the effects of soil texture, which involve root growth, moisture retention, and rate of leaching losses. It is impossible to say that in all environments clays are more valuable than sandy soils or *vice versa*. The lack of precise criteria imparts flexibility to the USDA system, enabling it to be adapted to local conditions, where particular properties are of special significance in rating land value; an example is the importance of soil-moisture retention in the climatic zone between the savanna and semi-arid environments.

It is not clear how the climatic limitation is to be interpreted in the USDA system, nor does it deal satisfactorily with wetlands. Wetness is classed on the basis of 'continuing limitations after drainage', using drainage measures considered practical at the present day. It is necessary to specify whether classification is on the basis of such drainage works as can be undertaken by the individual farmer, or whether a regional drainage scheme is under consideration. A special difficulty that arises in the humid tropics is that potential padi cultivation has totally different requirements from other forms of annual cropping. An important feature on which the system is not explicit is the extent to which economic considerations are taken into account. There is certainly some economic element, implied in the reference to 'practicable' measures for the removal of temporary limitations, and specifically noted in a reference to the need for a favourable input/output ratio to the farmer. Precise farm costing, however, is not attempted.

2   *Other systems*

The system of the U.S. Bureau of Reclamation (U.S. Department of Interior, 1953) is specifically for the classification of potentially irrigable land. It is again based on the limitations of the land rather than its positive qualities, but it contains precise local specifications for the permitted range of values within each land class. Costs of management and amelioration measures are taken into account. Land Classes 1–3 have progressively less repayment capacity for development under irrigation; Class 4 is special-purpose land;

Class 5 is non-arable at present but could become arable if some major works were undertaken; and Class 6 is non-arable.

The scheme of Haantjens (1965), designed for application to New Guinea, illustrates the results of modifying the USDA approach firstly by the incorporation of precise limiting values, and secondly by allotting different suitabilities for different types of land use. Haantjens rejects economic considerations on the grounds that they are ephemeral. Land is classified by an objective rating of the factors of the environment, as well as on a subjective, but defined, interpretation of these ratings in terms of land-use suitability. There are fourteen factors, including erodibility, stoniness, drainage, and nutrient status, which are each rated in up to seven classes, from $0 = $ best to 4, 5 or $6 = $ worst. Thus the erodibility rating is of the form:

$$0°-\ 2° \text{ slopes } = 0$$
$$2°-\ 6° \text{ slopes } = 1$$
$$6°-10° \text{ slopes } = 2$$

$$. \quad . \quad .$$

$$\text{over } 45° \text{ slopes } = 6$$

with a one-class adjustment on the basis of slope stability. Having obtained the fourteen ratings, each land unit is assessed according to its suitability for four types of use: annual crops (A), tree crops (T), improved pastures (P) and padi (R). For example a $6°-10°$ slope causes a substantial erosion hazard, or conservation costs, under annual cropping. This hazard is less serious for tree crops or pastures, but prevents padi cultivation, hence an erodibility rating of 2 gives suitability levels of A3, T2, P2, and R5. Suitabilities obtained from the fourteen factors are summed, by a prescribed method, to give overall suitabilities for each of the four types of land use. The overriding importance of one severe limitation is allowed for in the summation, in that the overall suitability can be no higher than the lowest indvidual suitability.

The separate evaluation for different purposes is an improvement on the USDA system. The relative desirability of each type of land use is not dictated; such priorities can be assessed subsequently, on economic or other grounds. In practice the suitabilities accord well with those obtainable by subjective assessment. But though subjective in construction, the scheme is objective in application; this has the advantage that once the ratings have been set up for a given area, they can be applied by different, including less skilled, personnel, with replicable results. As compared with the USDA system there is a gain in objectivity, at the expense of some loss in adaptability; the main advance, however, is the recognition that different types of land use may have markedly different environmental requirements.

An attempt has been made by Riquier, Bramao and Cornet, (1970) to devise a scheme based on positive land qualities, or productivity, rather than negative limitations. Values of nine soil parameters are set on a scale of 0–100, where 100 represents optimal conditions. These nine values are mathematically combined to give an overall rating ranging from 100 where all factors are optimal to zero where any one individual factor is zero. Like all schemes based on the addition or multiplication of numbers, this fails to take sufficient account of factor interactions, and of the different values of particular soil properties in different environments.

3   Systems used in Britain

In Britain there have been three national land classification schemes. The first was that of Stamp (1962), for which maps of the whole country at 1:625,000 were published. Three major groups, Good, Medium and Poor-quality land, were distinguished, divided into ten classes: for example Class 4, Good but heavy land; Class 9, Poor-quality light land. Although entirely subjective, the scheme was of much value in the war and post-war years when land-use planning first became significant.

Two further systems have since been concurrently applied. That of the Agricultural Land Service (1966, 1968) is frank about its purpose, which is 'for advising on the release of agricultural land for urban development'. It is a simple scheme of five grades, ranging from Grade I, Land with very minor or no physical limitations to agricultural use, to Grade V, Land with very severe limitations. Maps are produced as an overprint to the Ordance Survey series at 1:63,360.

The Land-Use Capability Classification of the U.K. Soil Survey (Bibby and Mackney, 1969) is designed to be produced in conjunction with soil survey maps. It is based on the USDA model but reduces the number of classes to seven by the omission of the USDA Class V (special-purpose lands including wetlands). Class definitions are similar to those of the USDA system. The division between arable and non-arable land again comes between classes 4 and 5. At the capability subclass level, five kinds of limitation are recognized: wetness (w), soil limitations (s), gradient (g), liability to erosion (e) and climatic limitations (c). Gradient is assessed not primarily with respect to erosion, but for its effects on mechanized farming. Guidelines for the permitted range of each limitation within each capability class are given; for example, land in Class 3 should have a gradient of less than 11° and a rooting depth of more than 25 cm. Unlike the position in most parts of the world, the land class becomes lower as rainfall increases—a good, if extreme, example to illustrate the changing significance of individual parameters in relation to the total environment. This scheme repays study, incorporating clarity of purpose, adaptation to local conditions, a combination of specific guidelines with flexibility of application, and simplicity and conciseness of presentation.

4   Systems incorporating economic factors

Economic considerations hover uneasily at the margin of the USDA and similar classifications, and it is unrealistic to suppose that they can be completely excluded from any land-evaluation scheme. Systems which apparently base evaluation on technology are in fact dependent on economics, for the cost of drainage, irrigation, soil conservation, fertilizers and other land-management operations in part dictate which technical measures are practicable.

There have been several economically-based procedures for land evaluation, notably the Cornell system (Conklin, 1959). In the U.K., attempts have been made to relate gross farm margins, from livestock enterprises as well as arable farming, to land class (Agricultural Land Service, 1966) and to soil series (Cruickshank and Armstrong, 1971). In a scheme outlined by Vink (1960), two groups of factors must be established, physical and non-physical. The former comprise climate, soil type, and—unlike most other systems—location. The latter are the economic situation, the technological situation and the management level. Given these, the first stage is to ascertain the crop rotation. For each year in the rotation estimates are made of crop yields, gross income, input costs (such as seed, fertilizers, labour) and hence net income. These are summed for the period of the rotation, and a deduction made for farm overhead costs. The result is the productivity, expressed in terms of net cash output per unit area. This can be converted into an index of relative productivity by expressing it as a percentage of some national or local base.

The focus of economically-based systems of land evaluation is net cash returns per acre. These vary, partly with costs, but mainly with crop yields. Unfortunately, resource surveys normally stop short at a qualitative assessment of land suitability for various crops, and do not attempt to predict yields. This is a serious hiatus in the logical sequence from environmental description to economic appraisal.

Reasons given by soil surveyors for not predicting crop yields are first, that the information is not available, and secondly, that yields in any case vary greatly with farming standards. Possible sources of information are first, yields under current farming; secondly, agronomic experimental work on type sites; and thirdly, comparison between soil characteristics and the physiological requirements of the crop. The first method suffices for developed countries; it requires, however, that an important part of soil survey procedure should be to visit farms for which yield data is available and examine the soils. Little or no time allowance for this essential activity is made in most soil survey specifications (Coulter, 1964). The second method, experimental work, needs a period of several years, longer for perennials than for annual crops. For under-developed countries, recourse has often to be made to the third method, some studies of which are noted below.

A prerequisite for yield prediction is to specify *levels of management.* For any given country this can be done in relation to current farming practices, and the following means of defining standard levels has been proposed (Steele, 1967):

Level 1: the most common existing cultivation practices
Level 2: superior existing cultivation practices
Level 3: optimum management

There is less difference between these levels in developed than in under-developed countries. In the latter, Level 2 refers to practices employed by more advanced farmers, following the advice of government agricultural extension services, but without complex techniques and with low capital investment. Level 3 refers to yields obtained on experimental stations or large farms, with access to advanced technology and with high capital investment.

IV Special-purpose and multi-purpose evaluation

A defect of general-purpose classifications is that different types of land use have different suitability requirements. Within the field of agricultural land classification alone this applies to individual crops. Thus grain crops tend to give higher yields on heavy-textured soils, and root crops on sandy soils. Groundnuts, with the special physiological requirement of peg penetration, grow best on sandy loams. In the climatic zone intermediate between savanna and rainforest, soil-moisture retention is critical for a perennial crop such as coffee, but less important for annuals. In pyrethrum cultivation, to give an extreme case, a period of chilling is necessary to initiate flower buds, and in the tropics yields increase with altitude. Various types of special-purpose evaluations are described in Stewart (1968).

1   *Single-crop evaluations*

In most agricultural textbooks the information on permissible and optimal environmental conditions for individual crops is notably vague, and hence gives little basis from which to estimate crop yields. Some studies have been made of individual crops, outstanding among which is work by Smyth (1966) on the selection of soils for cocoa. This was based on both empirical growing experience in West Africa and the physiological requirements of the crop for rooting, moisture and nutrients. The quality classification of potential cocoa soils given is:

Class I   Good
Class II   Fairly good
Class III   Poor
Class IV   Unsuitable.

Class III is divided into IIIn, d and p according to nutrient, drainage or other problems. Soil requirements for Classes I and II are:

Depth: > 150 cm
Drainage: free
Organic matter: > 3 per cent in upper 15 cm
pH: 6·0–7·5; no horizon $< 4·0$ or $> 8·0$
Cation exchange capacity: > 12 m.e./100 g in topsoil, > 5 m.e./
                                100 g in lower horizons
Exchangeable cations: in upper 15 cm, calcium $> 8·0$, mag-
                                nesium $> 2·0$ and potassium 0·24
                                m.e./100 g
Cation saturation: > 35 per cent in any horizon

Soil surveys are recommended on scales of 1:1,000,000 to locate likely areas, followed by 1:2,500–10,000 to indicate precise planting sites.

Other examples of one-crop investigations are studies of coconuts on Christmas Island (Jenkin and Foale, 1968) and oil palm in Gambia (Hill, 1969). The former, although covering only a small area, is noteworthy in combining a detailed discussion of environmental conditions with a thorough treatment of agronomy and management aspects.

The latter could be paraphrased 'If you want to grow oil palm, don't do so in Gambia'; it is none the less a comprehensive study, taking the ecological requirements of the plant as a basis.

The main user-requirement in arable land evaluation is crop yields, and it is precisely on this point that there is little reliable information. A comparable situation exists in surveys for grazing purposes. Here the critical parameter is the *carrying capacity*, defined as the level of livestock use which a pasture will tolerate without permanent deterioration. Where well-managed ranching exists, this can be estimated by comparative methods (Condon, 1968). The normal situation in under-developed countries, however, is that over-stocking, with vegetation degeneration, has already occurred (Ch. 11, II, 2). Estimation of the reduction in livestock numbers necessary to restore a steady state is then difficult, and no techniques have been developed. Thus in a comprehensive ecological survey of Western Zambia by Verboom (1965; Verboom and Brunt, 1970) the estimates of carrying capacity are stated to be entirely subjective.

2   *Evaluation for non-agricultural uses*

Special-purpose evaluations outside the field of agriculture include appraisals for engineering and military purposes. The main engineering consideration in rural areas is road construction. In rich, densely settled countries, road alignments are dictated largely by considerations of the existing economic infrastructure. Foundation properties are studied by taking numerous samples along all possible alignments. In under-developed countries, and sparsely settled rich lands, such methods are inefficient. Relatively rapid and inexpensive procedures for selecting alignments were first developed in South Africa (Brink and Williams, 1964) and have since been used by the Tropical Division of the U.K. Road Research Laboratory (Clare and Beaven, 1965; Beaven, 1966; Dowling, 1968).

The method commences with a land-systems survey by air photograph interpretation, which includes mapping down to land facet level. This map is used first to eliminate from further consideration those land systems and facets that are clearly unsuited to road construction, and secondly as a basis for stratified sampling to test properties. The properties relevant to road alignment fall into four groups:

(i) *Landforms:* Slope steepness and relief pattern (arrangement, orientation and continuity of slopes), which affect alignment, length of route and amounts of cut-and-fill; and drainage density, pattern, and stream-flow regimes, which affect bridge construction and maintenance.

(ii) *General properties of the regolith* (*soil* in the engineering sense): depth of weathering, surface and subsurface drainage, as related to foundation requirements and verge erosion.

(iii) *Engineering parameters of the regolith:* including Atterberg limits, plasticity index, bulk density, compaction properties, linear shrinkage and particle-size distribution.

(iv) *Availability and nature of constructional materials:* crushable rock, gravel, sand and laterite.

Military land evaluation is for the purpose of predicting military requirements, primarily ease of vehicle movement away from roads, in territory to which there is no field access. The system developed by the Military Engineering Experiment Establishment in Britain is based on the principle of identifying land systems in enemy territory similar to those to which there is field access. Comprehensive tests have achieved a satisfactory level of prediction (Miller, 1967, and references cited therein).

3   *Multi-purpose evaluation systems*

Multi-purpose evaluation is the combination of several special-purpose evaluations, with or without a system for the allocation of priorities. It may be illustrated by two examples.

(*a*) *Malaysia.* The Economic Planning Unit of the Government of Malaysia has carried out a programme of land inventory designed to be applied at the national planning level (Panton, 1970). Six types of data are obtained: present land use, land tenure, mineral potentiality, soil suitability for agriculture, forest productivity, and water resources. These are combined to give a simple five-class ordering of value:

Class I	Land with high potential for mineral development
Class II	Land with high potential for agriculture
Class III	Land with moderate potential for agriculture
Class IV	Land with high potential for forestry
Class V	Land with little mineral, agricultural or productive forestry potential, but suitable for protective forest reserves, water catchments, game reserves or recreation

This method of ordering is indicative of the priorities in under-developed countries. Mineral exploitation has absolute priority over all other uses, even at the expense of destroying good agricultural land, because of its much higher economic productivity per unit area. Agriculture takes precedence over forestry on similar grounds, and recreational use comes low in the order.

(*b*) *Canada.* The most comprehensive national scheme of land evaluation is the Canada Land Inventory (1969–70). The programme, initiated by the Agricultural Rehabilitation and Development Act of 1961, involves the co-operation of over 100 federal and provincial agencies. It is intended to cover 2,500,000 sq km and to produce 15,000 map sheets. Because of the scale of the operation it was decided at an early stage to use a system of computerized data storage and retrieval (Tomlinson, 1968).

The following separate maps are produced at a scale of 1:250,000:

(i)   Soil capability for agriculture
(ii) Land capability for forestry
(iii) Land capability for recreation
(iv) Land capability for wildlife—ungulates
(v)   Land capability for wildlife—waterfowl
(vi) Land capability analysis

The first three are the major competitive demands for land in the rural environment. The fourth and fifth are requirements of special significance to Canada. Each of the five special-purpose maps is based on similar principles to the USDA approach, with seven classes in order of suitability for the purposes concerned, and fairly large numbers of subclasses indicating the types of limitations. For example in the recreation map, Classes 1–3 have capabilities for outdoor recreational uses of an intensive nature, Classes 4–6 for uses based on dispersed activities, and Class 7 has very low recreational capabilities; Subclass B is land adjacent to angling waters, and Subclass G a significant glacier view or experience.

The Land Capability Analysis is produced from the special-purposes maps by an overlay process. The first three classes from each purpose are initially considered; where none is present, Class 4 areas for agriculture and forestry are added, and subsequently perennial forage (agriculture Class 5). Any remaining areas are classified as low-potential land. The prime capability classes for each purpose are considered to be of equal value. Where conflict exists between two or more purposes at the same capability level, these are resolved in committee. The map is intended as a guide to, and not a substitute for, land-use planning. The map omits considerations of the local demand for, and supply of, land of a particular type, whilst recognizing that these will be important at the planning stage.

Multi-purpose evaluations involve a central problem in land evaluation, that of deciding between competing uses of a differing nature, which employ different criteria to assess suitability. The only common means yet devised is to reduce all evaluations to economic terms, and employ cost/benefit analysis; but the procedures for the benefit assessment of land in which conservationist aspects occur is far from satisfactory. Partial resolution can sometimes be made in terms of multiple uses. Thus agriculture is consistent with low-density recreational use, but not with the larger numbers of people that can reach countryside in societies with mass car-ownership. Forestry can tolerate higher recreational densities. Wildlife conservation has only limited compatibility with other uses, and, owing to the small numbers of people who patronize it, is often difficult to justify in economic terms.

V An example: land evaluation in a rainforest environment

The interaction of environmental, technological and economic considerations may be illustrated by a hypothetical example of stages in evaluation for a land-settlement project. It is based on actual conditions on a settlement scheme in Malaya planned in 1965–6 (Tippetts *et al.*, 1967), simplified to clarify the issues.

I   *Description*

It will be assumed that the area is situated at a low altitude in the tropical evergreen rain forest zone, and is at present covered by the natural forest. It has a mean annual rainfall of 2,500 mm, evenly distributed throughout the year. Within the area considered, the variations in climate are not sufficient to be of significance in land potential. Internal differentiation in land potential is therefore caused by landforms and soils and their secondary effects. Following natural resource survey, maps of these two factors are therefore the basic documents for land evaluation.

The landform and soil maps are shown in Figs. 1.1A and 1.1B. A stream with a level floodplain flows across the area, and there is a hill to the north-east. An area of gentle (0°–5°) slopes lie on one side of the valley, with moderate (0°–10°) slopes adjacent to it. The floodplain is subject to frequent inundation, and has hydromorphic alluvial soils. The upper part of the hill has shallow and stony soils, the lower part colluvial sandy soil of 30–50 cm depth. There are three other soil-mapping units, all with ferallitic soils but varying in texture and depth. These are (i) a red, deep, clay; (ii) a yellow sandy soil averaging 100 cm depth; (iii) an area which soil survey failed to map

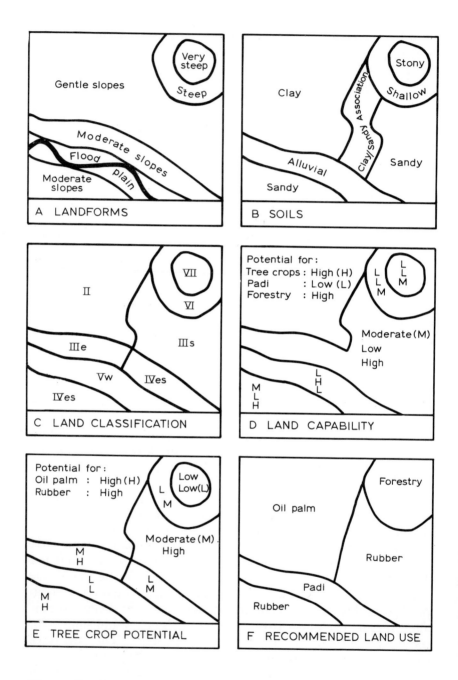

Fig. 1.1 Land evaluation in a rainforest environment

at series level, shown as an association of these two soils. Doubtless these are caused by differences in the parent materials of shale, sandstone, and intercalated beds of the two.

2 *Appraisal*

Fig. 1.1C shows this area classified according to the USDA system. There is no land that is both level and free from flooding, and therefore no land of Class I. The best land, in Class II, is the red clay on gentle slopes. Either moderate slopes or an effective depth of 100 cm is considered sufficient to lower the land to Class III, Subclasses IIIe and IIIs respectively; both these conditions together lower it to Class IV. The valley-floor wetland is in the special Class V, and limitations of both soil and slope place the hill into low classes.

A general-purpose classification such as this serves a purpose at the regional planning scale; for example, the proportion of the total area in capability classes I-III forms a guide in selecting regions for possible development and thus, in the first instance, for more detailed resource survey. At the local scale level of this example, however, it is inadequate for the main planning purpose, that of providing land-use recommendations. For this, the resource data is reassessed in terms of three types of land use, namely tree crops, padi cultivation and commercial forestry. Two other uses are not considered: annual crops, because of the severe erosion hazard, and grazing, because of lack of technical knowledge about its potential in this environment. Three maps of the capability for these uses are produced, and a fourth by superimposition, shown as Fig. 1.1D. Thus the potential for forestry is low in the valley floor, because of difficulties of mechanical extraction, but moderate or high elsewhere. Padi capability is the reverse. On the basis of any one of these maps, myopic land-use recommendations could be made. Thus the forester might suggest a pattern of forestry over most of the area, with non-forest uses in the valley floor, while a Malay villager might propose the opposite.

The multi-purpose capability map does, however, permit certain decisions to be taken. A *low* suitability implies that there is no potential for the use concerned: hence the valley floor and the hill do not have competing uses, and can be assigned to padi and forestry respectively. The greater part of the land could support two alternative and non-compatible uses, tree crops and forestry. At this point economics is brought in. The long-term economic productivity per unit area, and the return to investment, is greater from tree crops than from forestry. Hence where environment permits two uses, economics dictates that tree crops should take precedence.

Possible tree crops for the area are bananas, oil palm, and rubber. Bananas are ruled out by uncertainties over the long-term market price, and by the fact that special handling facilities would have to be constructed, which the present scheme is not sufficiently large to warrant. A further map, Fig. 1.1E, is therefore produced, showing qualitatively the potential for oil palm and for rubber. Oil palm is a highly productive crop, in terms of calories produced per unit area, but has correspondingly high demands for nutrients if yields are to be sustained. Under the high rainfall, nutrients added in fertilizers will be removed by leaching in all soils, but more rapidly on the sandy soil, both because of its higher permeability and lower cation exchange capacity. Moreover, it is a new crop to Malaya, and the lack of agronomic information based on experience of cultivation imparts a note of caution to the making of recommendations for its cultivation, in view of the high capital investment involved. Rubber, whilst it too grows best on deep, well-structured soils, is less demanding, and is known from elsewhere to yield satisfactorily on shallower and sandy soils. Hence, whilst the red clay on gentle slopes or Class II land has a high potential for both crops, the other tree-crop land is rated more suitable for rubber than for oil palm.

This consideration is counteracted by present prices and market prospects. The net profit per unit area, per unit of labour, the gain to the national economy and the returns on investment are all higher for oil palm than for rubber. Moreover the market in vegetable oils is steady, by comparison with that in many primary products, whereas rubber faces competition from synthetics which, despite a continuing demand for natural rubber, depresses the price. So the economic pressure is to plant oil palm over the whole of the cultivable area.

3   Development

These conflicting pressures are resolved in committee, and Fig. 1.1F shows a possible outcome, in which the potential has been translated into land-use recommendations. Forestry and padi occupy the land to which they alone are suited and, in the solution shown, a cautious recommendation has been made, restricting oil palm to the best land. A further aspect that has entered at this stage is that several of the mapping units on Fig. 1.1E, although representing the realities of the environment, are too small to serve as practical farming units for different crops. Hence oil palm is carried down to the floodplain across the steeper valley side, whilst the boundary kink that originated in the soil map (perhaps caused by a geological fault) and appeared on all evaluations has disappeared under the thick felt pen of the planner.

VI Conclusion

Three stages can be identified in land appraisal: qualitative, quantitative and economic evaluations; and these are illustrated in Table 1.1.

The stages differ according to the types of information employed, the scale of mapping, and the types of planning decision for which they provide information. Qualitative evaluation employs mainly environmental information, linked to a general consideration of technological aspects; economics is largely subsumed in the latter. It is useful at the reconnaissance stage of survey, for guidelines to planning on a national or major regional scale. Quantitative evaluation requires that technological information (for example on agronomy, or wildlife habits) be incorporated in detail, together with more specific information on those environmental factors that are relevant to given uses. Information drawn from economics is used to rule out certain types of clearly unprofitable use. This stage enables land-potential maps to be produced at either semi-detailed or detailed scales, depending on the intensity of survey. Land-planning recommendations, and in particular investment decisions, require the further stage of economic appraisal, taking the results of the quantitative evaluation as a basis. Other planning considerations, for example conservation and matters of political expediency, may yet override the economic evaluation before it reaches implementation.

The subject of rural land evaluation is currently in an active stage of evolution; an international conference is being held by the F.A.O. whilst this book is in the press. Judgements on the merits of existing systems are therefore likely to be rendered obsolete. Some general comments may be offered. Land evaluation is done for practical reasons, and hence there is an all-important need to define the purpose of any given project. This in no way implies that the purpose must always relate to immediate planning decisions; for example, a long-term assessment of a country's ultimate resource potential is a valid aim. General-purpose classifications that rate land in a single order of value (I, II, III . . .) have inherent limitations, in that the technical requirements of different types of land use, and even different forms of agricultural use, do not correlate closely. It is impossible to produce definitive or standardized evaluation maps, except in general terms. Requirements of technology and economic conditions are constantly changing, and there is no escaping a continuing reappraisal, based on an understanding of the environment. The importance of studying technology in detail is generally under-rated, by comparison with the recognized value of understanding the environment and economics.

Table 1.1 Classification of purposes and stages in rural land evaluation

Purpose	Qualitative evaluation	Quantitative evaluation	Economic evaluation
Agriculture, arable	Suitability for—individual crops —annuals, tree crops —arable use	Crop yields, under defined management levels	Annual income (gross and net) from crop production ; present worth of annual values (using discounted cash flow) ; cost benefit analysis of investment
Pastoralism	Suitability for—grazing of unimproved pasture —pasture improvement	Stock carrying capacity	Annual income from livestock production
Forestry	Suitability for—logging —forest plantations	Timber resources Annual growth rates	Income from logging Predicted annual income from forestry
Irrigation	Suitability of land for irrigated agriculture	Crop yields, with and without irrigation	Annual income with and without irrigation, in relation to costs of irrigation ; cost/benefit analysis of investment
Primary production from land resources	General-purpose land capability classification Suitabilities for annual crops, tree crops, grazing, forestry, irrigation	—	—
Engineering	Suitability for roads and other engineering works	Parameters relevant to road construction and maintenance	Estimates of income from each type of land use
Recreation	Suitability for recreation : high-intensity, low-intensity uses	Estimates of output from each type of land use	Costs of road construction and maintenance ; benefits Economic evaluation of predicted recreational use
Conservation	Wildlife sites, importance Historical sites, importance	Density and frequency of acceptable use Rarity of site, species	—
Multi-purpose evaluation	Relative suitabilities for different types of land use	Ordering of relative suitabilities in accordance with defined criteria	Ordering of suitabilities for different purposes in accordance with assigned economic values

There is a need in under-developed countries to devise better methods for prediction of crop yields, stock-carrying capacities, and other quantitative assessments of land productivity. Present procedures for resource survey need modifying, to incorporate activities designed specifically for such quantitative assessments (Young, 1973). For planning decisions in developed countries, present methods of economic evaluation are unsatisfactory when applied to classes of land use outside those involving the production of goods—for example, to landscape and wildlife conservation or to recreation.

Land evaluation is a process involving the gathering of information of widely differing types (environmental, technological and economic), its integration in a spatial context, and the directing of this synthesis towards specified purposes, namely those of land use and its planning. Gibbons, Rowan and Downes (1968), in a reaction against the tendency to overvalue automated methods, plead for the 'role of humans in land evaluation'. Routine operations, of survey, storage and data analysis, can be left to instruments. The special role of people is to identify a purpose, direct automata towards the operations necessary to fulfil it, and to take decisions based on the outcome. If there is any one viewpoint from which it is easiest to start in order to gain an appreciation of the whole, it is probably from the central position, that of technology. This is perhaps why farmers make such good land-capability assessments, albeit from a restricted attitude as regards aims. But there is no reason why a successful synthesis should not be achieved from the starting points of either natural science or economics. Rural land evaluation provides a particularly clear illustration of the fact that real-world situations involve the interaction of different classes of phenomena, in what would scientifically be called an uncontrolled situation. The approach to problems of this nature must be problem-orientated, starting from a definition of aims and proceeding into whatever subject-matter is necessary for their accomplishment.

References

AGRICULTURAL LAND SERVICE (1966): Agricultural land classification. *Ministry of Agriculture, Food and Fisheries, Technical Report* **11**. London.
AGRICULTURAL LAND SERVICE (1968): *Agricultural land classification map of England and Wales. Explanatory note.* Pinner.
BEAVEN, P. J. (1966): Roadmaking materials in Basutoland. Classification of soils. *Road Research Laboratory Report* **47**. Crowthorne.
BIBBY, J. S. and D. MACKNEY (1969): Land use capability classification. *Soil Survey, Technical Monograph* **1**. Rothamsted.

BRINK, A. B. and A. A. B. WILLIAMS (1964): Soil engineering mapping for roads in South Africa. *C.S.I.R. Research Report* **227.** Pretoria.

BRINK, A. B., J. A. MABBUTT, R. WEBSTER and P. H. T. BECKETT (1966): Report of the working group on land classification and data storage. *M.E.X.E. report* **940.** Christchurch, Hants.

C.S.I.R.O. (1953–71): *Land Research Series* **1–29.** Melbourne.

CANADA LAND INVENTORY (1969a): Soil capability classification for agriculture. *Report* **2.** Ottawa.

(1969b): Land capability classification for outdoor recreation. *Report* **6.** Ottawa.

(1969c): Land capability classification for wildlife. *Report* **7.** Ottawa.

(1970a): The Canada land inventory: objectives, scope and organization. *Report* **1,** second edition. Ottawa.

(1970b): Land capability classification for forestry. *Report* **4.** Ottawa.

CHARTER, C. F. (1957): The aims and objects of tropical soil surveys. *Soils and Fertilizers* **20,** 127–8.

CHRISTIAN, C. S. and G. A. STEWART (1953): General report on survey of Katherine–Darwin region, 1946. *C.S.I.R.O. Land Research Series* **1.** Melbourne.

CLARE, K. E. and P. J. BEAVEN (1965): Roadmaking materials in northern Borneo. *Road Research Laboratory Technical Paper* **68.** Crowthorne.

CONDON, R. W. (1968): Estimation of grazing capacity on arid grazing lands, *in* G. A. STEWART, editor: *Land evaluation.* Melbourne.

CONKLIN, H. E. (1959): The Cornell system of land classification. *Journal of Farm Economics* **41,** 548–57.

COULTER, J. K. (1964): Soil surveys and their application in tropical agriculture. *Tropical Agriculture* **41,** 185–96.

CRUICKSHANK, J. G. and W. J. ARMSTRONG (1971): Soil and agricultural land classification in County Londonderry. *Transactions of the Institute of British Geographers* **53,** 79–94.

DAVIDSON, B. R. (1965): *The northern myth: a study of the physical and economic limits to agricultural and pastoral development in tropical Australia.* Melbourne.

DOWLING, J. W. F. (1968): Land evaluation for engineering purposes in northern Nigeria, *in* G. A. STEWART, editor: *Land evaluation.* Melbourne.

FARMER, B. H. (1971). The environmental sciences and economic development. *Journal of Development Studies* **7,** 257–69.

GIBBONS, F. R., J. N. ROWAN and R. G. DOWNES (1968). The role of humans in land evaluation, *in* G. A. STEWART, editor: *Land evaluation.* Melbourne.

HAANTJENS, H. A. (1965): Agricultural land classification for New Guinea land resources surveys. *C.S.I.R.O. Division of Land Research Technical Memorandum* **65/8.** Canberra.

HILL, I. D. (1969): An assessment of the possibilities of oil palm cultivation in Western Division, The Gambia. *Directorate of Overseas Surveys, Land Resource Study* **6.** Tolworth.

JENKIN, R. N. and M. A. FOALE (1968): An investigation of the coconut-growing potential of Christmas Island, vols. 1 and 2. *Directorate of Overseas Surveys, Land Resource Study* **4.** Tolworth.

KLINGEBIEL, A. A. and P. H. MONTGOMERY (1961): Land-capability classification. *Soil Conservation Service, Agriculture Handbook* **210.** Washington D.C.

MABBUTT, J. A. (1968): Review of concepts of land classification, *in* G. A. STEWART, editor: *Land evaluation*. Melbourne.

MILLER, T. G. (1967): Recent studies in military geography. *Geographical Journal* **133**, 354–6.

MOSS, R. P. (1968): Land use, vegetation and soil factors in south-west Nigeria. A new approach. *Pacific Viewpoint* **9**, 107–27.
(1969): The appraisal of land resources in tropical Africa. A critique of some concepts. *Pacific Viewpoint* **10**, 18–27.

NYE, P. H. and D. J. GREENLAND (1960): *The soil under shifting cultivation*. Harpenden.

PANTON, W. P. (1970): The application of land use and natural resource surveys to national planning, *in* I. H. COX, editor: *New possibilities and techniques for land use and related surveys*, 129–38. Berkhamsted.

RIQUIER, J., D. L. BRAMAO and J. P. CORNET (1970): A new system of soil appraisal in terms of actual and potential productivity. *F.A.O. Report, A.G.L.: T.E.S.R./70/6*. Rome.

ROBERTSON, V. C., T. N. JEWITT, A. P. S. FORBES and R. LAW (1968): The assessment of land quality for primary production, *in* G. A. STEWART, editor: *Land evaluation*. Melbourne.

SMYTH, A. J. (1966): The selection of soils for cocoa. *F.A.O. Soils Bulletin* **5**. Rome.

STAMP, L. D. (1962): *The land of Britain: its use and misuse*. London.

STEELE, J. G. (1967): Soil survey interpretation and its use. *F.A.O. Soils Bulletin* **8**. Rome.

STEWART, G. A., editor (1968): *Land evaluation*. Melbourne.

TIPPETTS-ABBETT-MCCARTHY-STRATTON and HUNTING TECHNICAL SERVICES LTD. (1967): *The Jengka Triangle report*, 4 vols. Kuala Lumpur.

TOMLINSON, R. F. (1968): A geographic information system for regional planning, *in* G. A. STEWART, editor: *Land evaluation*. Melbourne.

U.S. DEPARTMENT OF INTERIOR (1953): *Bureau of reclamation manual, Vol. V, Irrigated land use, Part 2, Land classification*. Washington D.C.

VERBOOM, W. C. (1965): The Barotseland ecological survey, 1964. *Internal Report, Department of Agriculture, Zambia*. Lusaka.

VERBOOM, W. C. and M. A. BRUNT (1970): An ecological survey of Western Province, Zambia, with special reference to the fodder resources, Vols. I and II. *Directorate of Overseas Surveys, Land Resource Study* **8**. Tolworth.

VINK, A. P. A. (1960): Quantitative aspects of land classification. *Transactions of the 7th International Congress of Soil Science* **5**, 371–8.

YOUNG, A. (1973): Soil survey procedures in land development planning. *Geographical Journal* **139**, *in press*.

2 Agriculture and agrarian land planning

P. T. Wheeler

I Perspective on the geographer's role

Agriculture is far more than the mere 'cultivation of the soil' (*Concise Oxford Dictionary*); it is the modification of the natural land environment for the production of plant and animal crops for subsistence and exchange. This is an activity which has occupied the greater part of humanity for millennia, and upon which over half of the world's population still depends. (Although the percentage of the world's population directly connected with agriculture has fallen from the 62 per cent of 1937, it remains above 50 per cent (U.N.F.A.O., 1959; 1970).) Societies must have access to adequate food supplies for survival, and the continued existence of most states has largely depended upon a successful agriculture, while many have declined with the decay of their agriculture. Developments in advanced economies, whereby a country such as the United Kingdom may have less than three per cent of its people engaged in agriculture, have hitherto been rare indeed, though they are likely to become more common. Yet agriculture, especially technologically simple agriculture, is very strongly influenced by the natural environment as well as by the abilities and preferences of farmers and by historical accident, so that its characteristics, function and products may vary widely from place to place and from time to time.

Agriculture is, therefore, a major human activity, of long history and of worldwide occurrence, which yet shows marked variations in its characteristics and inequalities in the distribution of its enterprises. The agrarian landscape is the visible expression of the practice of this, man's oldest and largest industry, and the analysis of the agrarian landscape is a legitimate and fascinating task for the geographer. However, the materials he will have to handle in order to do this are extraordinarily complex. He may wish specifically to examine past agricultural patterns, in which case he will need the skills of the historian and possibly of the archaeologist, but the odds are that even if he intends to restrict himself to the present he will still be forced to look back to the past to explain the origins of current patterns and to understand existing systems. Certainly such an understanding is essential to the successful prediction of the effects upon agriculture and the agrarian landscape of such developments as extreme urbanization, industrial and capital intensive agriculture and amenity pressures upon rural areas.

1   *Approaches to the study of agrarian landscapes*

Agriculture may be studied in a variety of ways, depending on the training and interests of the person concerned. The geographer is interested in the elucidation and explanation of distributions of crops, of stock, of agricultural practices and organization, and of the exchange of agricultural products (see also the discussion on marketing in Chapter 6). His characteristic tool is, of course, the map. In this case it will almost certainly be primarily the land-use map, supplemented by a variety of other maps such as those dealing with farm types and organization and with individual stock and crop distributions. But since 'land use patterns can only be understood by understanding the relevant processes of human decision making' (Found, 1971, p. 165), the geographer will need to draw not only upon his own observations but also upon the work of many other specialists, so that he may be able to recognize those processes which have led to the mapped distributions.

The investigation of agriculture, therefore, is likely to be interdisciplinary, and Young (Ch. 1) indicates that this is sometimes through a team activity. In the past such work, individual or cooperative, was generally *pure* in the sense that it was concerned only with description of the past and present and with the establishment of principles to explain what was described. Now, however, two factors have induced a move towards *applied* research, i.e. towards the application of abstract principles to actual society in order to produce certain desired changes. In other words, research has become linked with both agricultural and economic planning policies and methods. The first factor is the increasing expense and complexity of team research programmes, which means that often only governments and official bodies can afford to sponsor them. The second factor is the increasing awareness on the part of national governments that some degree of planning is necessary if their societies are to be adjusted to the modern world. Since in most countries agriculture is still the basic social and economic activity of the majority of the population, this is where investigation and planning must often be concentrated and where there will, therefore, be a call for research. The individual geographer may still be able to carry on his work by himself, but his chances of applying the results of his research are likely to be small unless he has such official backing.

The relevance of work in physical geography to the planning and development of rural areas is shown elsewhere in this volume (Ch. 1). In the present chapter the emphasis will be on the contribution which can be made by the geographer approaching this subject from the human—that is to say the historical, economic and social—point of view.

The study of agrarian landscapes involves three principal stages:

(i)   data collection
(ii)   data explanation
(iii) prediction based on recognized trends, including the prediction of what will happen if there is interference with the present trends.

This chapter can do no more than touch upon some of the problems that arise under these headings; other problems are discussed by Gregor (1970). The chapter is concerned to a great extent with the problems of data collection, for it can be argued that the geographer's chief role to date has been in the collection of basic information upon which data explanation and prediction of future events may be based. Unfortunately, as will be shown by reference to two case studies drawn from Britain and Nigeria respectively, the task of data collection is by no means easy. However, accurate data collection is vital if the planning of agriculture, or of the agrarian landscape as a whole, is to be either meaningful or sensible, and it requires all the expertise which, by virtue of his training, the geographer has at his command.

II Data collection: establishing the facts

1 *Published data sources in Britain*

The first question to ask in this connection is 'What are the facts required for a geographical study of agriculture?', and the initial answer must be 'The nature and quantity of crops and stock raised, their location and organization'. The geographer would probably begin his search for such information by looking at the official published data, and of these there are normally large quantities in advanced countries. In Britain a primary source is the Ministry of Agriculture's annual census of agriculture, mainly based on the 4 June farm returns. The latest volume at the time of writing, *Agricultural Statistics 1968/1969 England and Wales* (Ministry of Agriculture, 1970; Interdepartmental Committee on Social and Economic Research, 1958) for instance, contains information on:

(i) the area under crops and rough grazing (41 categories) and the numbers of livestock (48 categories) by county
(ii) the production by acreage, yield per acre and total output of 15 crops
(iii) the number, size and distribution of farm holdings and the frequency distribution of crops and livestock by size of farm
(iv) the number of agricultural workers by county, and the total numbers of selected types of farm machinery
(v) a special section devoted to horticulture
(vi) a section on prices.

(A similar volume is available for Scotland by the Department of Agriculture; Northern Ireland is dealt with in the material consolidated by the Ministry for the United Kingdom as a whole.) Such statistics can be traced back for over a hundred years to 1866, though the earlier series are less elaborate and less definitive (Best, 1959; Best and Coppock, 1962; Coppock, 1956; Ministry of Agriculture, 1969). They can be supplemented from many published sources. Official sources include parliamentary reports and the many other publications of the Ministry of Agriculture. There are also semi-official sources like the various studies issued, for example, by the regional colleges of agriculture, or the volumes of such bodies as the Royal Agricultural Society of England, and of landowners' and farmers' associations. Other published materials include such individual studies as Ernle's *History of English Farming* (1961) and that by Symon (1959) relating to Scotland. This wealth of literature can be followed back to the county descriptions of the first Board of

Agriculture during the Napoleonic wars (Mitchison, 1962) and even beyond, though increasingly sporadic and esoteric, to the Domesday Book, that incomparable datum for historians, economic historians and geographers.

But although these riches are undeniable and still invite much assiduous exploration, there are deficiencies which irritate the agricultural geographer but also goad him to further action. To begin with, official statistics are usually published for the country as a whole or by county, and in a land as varied as Britain, these units do not give a network fine enough for adequate distributional analysis. True, parish statistics can be obtained if the scholar is able either to pay for their abstraction or to go to the Ministry's library at Guildford (or Edinburgh for Scotland). But even so, he will meet difficulties. For instance, he may only have parish statistics provided that it remains impossible to identify any single farm within that parish; this safeguards the confidentiality of individual returns, but complicates research. Furthermore, parishes vary greatly in size and shape, while the individual parish itself tends often to unite land of contrasting physical endowment rather than to cover a homogeneous area. Again, and especially with modern amalgamation of holdings, farms frequently transgress administrative boundaries, whereas returns are made as from one location, usually the farmstead. In this way a parish may be credited with lands lying in neighbouring parishes, and sudden changes of acreage and livestock in parish totals may well be due to changes in farm boundaries rather than to real changes in the agricultural system. Of course the parish and other returns can be used, as Coppock (1964b; 1971) and many others have demonstrated; but by and large, if the geographer wishes to establish precise agricultural distributions and the actual use of the land, he must do so himself, often by himself, in the field.

When he starts doing this, he will realize that some vital questions are dealt with inadequately or not at all in the official sources, and that other sources are either local in their coverage or equally inadequate if universal. For example, the size, shape and layout of the field, the farmstead or the whole farm are nowhere properly described or analysed, although individual studies (Brunskill, 1970; Coppock, 1971; Harvey, 1970; Simmons, 1964) reveal enormous local variations and intricate historical causation; Stolper's *Planning without facts* (1966) might almost have been written of the British Ministry of Agriculture. The only official survey of these matters, the wartime National Farm Survey (Ministry of Agriculture, 1946), was never fully published, and the original documents may have been partly lost.

An even worse gap in British land-use records is created by the absence of any one authority responsible for making a universal record of the use of land in Britain. The Ordnance Survey can provide accurately measured areas, mainly at field or administrative unit level, but it does not analyse use; the Ministry of Agriculture restricts itself to land used directly for agriculture; the Forestry Commission records its own properties and carries out occasional national censuses of woodlands and of hedgerows and parkland timber; but the use of built-up and other developed land, essential as knowledge of this may be for an understanding of the pressure upon agriculture, is in effect covered only by the disparate documents of varying age compiled (but not necessarily published) by local planning authorities. Even then, land held by the military may be virtually omitted. In fact, the only universal coverage of Britain was that of the Land Utilization Survey of Britain, carried out in the 1930s under the direction of Sir L. Dudley Stamp, publication of which was only completed in 1947; while the neglect of the more recent survey directed by Miss Alice Coleman and published at the 1:25,000 scale remains distressing in a country paying much lip-service to physical planning (Coleman and Maggs, 1968).

2   *Case study one—the Crofting Survey*
What, then, may happen when the geographer does take to the field to establish the facts? Every piece of research has its own story. but let us here take the example of the Crofting Survey. This was undertaken in the Highlands of Scotland, which, largely owing to their inaccessibility and poverty of natural endowment, have for more than two hundred years represented a *depressed* area within Great Britain.

(*a*) *Definition of the problem.* A croft is, broadly speaking, a tenant holding, in the seven Crofting Counties of Scotland (Argyll, Inverness, Ross and Cromarty, Sutherland, Caithness, Orkney and Shetland), of less than twenty hectares arable and/or of less than £50 annual rent, that has been registered as a croft. In 1962 it was calculated that there were about 16,000 effective crofting units, supporting approximately a quarter of the population of the Crofting Counties, which at that time were receiving more than £30,000,000 in annual grants and subsidies, a sum that must since have greatly increased (Moisley, 1962b; Financial Times, 1961). Crofts, however, tend usually to be grouped on one of the limited patches of good land in the Highlands. A ring-fence or *head-dyke* will surround this area of *inbye*, within which the units are most often individually fenced and operated. But the most valuable land, taken as a whole, usually lies beyond the head-dyke in the *outrun* grazing land held in common. This is managed by the elected Common Grazings Committee, and possession of grazing rights implies membership of the community or *township*. However, since crofts (even with their grazing rights) are by definition small, a supplementary source of income is normally necessary. This leads to depression in the Crofting Counties, for local alternative employment is strictly limited, and emigration ensues, which in turn leads to neglect of the land and (especially in a partly co-operative community) the impoverishment of local life.

(*b*) *Method of enquiry.* Many official enquiries have been held on this area of persistent economic depression upon which so much money has been spent and about which so many individuals have published work. But when, in 1956, the protests resulting from a proposal to dispossess large numbers of crofters in North and South Uist and Benbecula in order to establish a rocket range and allied military installations provoked Moisley and Caird of the Department of Geography, University of Glasgow, to undertake a detailed investigation of the agricultural system under threat, a chief result was the revelation of how little was known of the geographical and related economic and social facts which made that system work. A programme of work developed, growing into studies of Benbecula and South Uist in 1956, Barra in 1957 and of parts of Harris and of Vaternish (Skye) in 1958. These were carried out with the assistance of parties of students from Glasgow; but in 1958 another study of Park in Lewis was carried out in co-operation with a party from the Geographical Field Group. Work, again with the Geographical Field Group, followed in the Uig and Point districts of Lewis in 1959, and in 1960 a study of the island of Unst in the Shetlands was undertaken by the Geographical Field Group alone. Similarly, a series of studies of Orkney by students of the Ponteland College of Education, Northumberland, culminated in an investigation of much of the West mainland of Orkney by the Geographical Field Group in 1968. During this period investigations at various other hands have become far more common in the Crofting Counties, and the official attitudes have changed from grudging tolerance to co-operation and encouragement as the value of basic geographical research in the formulation of planning policies has been demonstrated.

The sheer organizational difficulties of collecting such information should not be overlooked. Apart from preliminary reconnaissance and archive work, research in the Crofting Counties involved informing: the Church of Scotland Minister (or the Roman Catholic priest in some of the southern Hebrides, or the Free Church Minister in some other areas) as a critical leader of local opinion; the Secretary of the Common Grazings Committee, as a key crofter; the local office of the Department of Agriculture for Scotland, as a source of current official data; the Crofters Commission, as the source of official registers and lists of crofters; and the local landowner, as the source of estate archive material. It proved in practice possible to carry out an average of five interviews, plus related land-use survey, type-of-building analysis, etc., per worker per day. This meant that one could calculate the length of time to be spent upon a given field-work programme. For example, 120 crofts could be enumerated in four weeks or 24 working days by a single worker, but in two working days by a team of a dozen people. On the other hand, accommodating a dozen people can strain the resources of an isolated and small community. Furthermore, since the co-operation of the local people is essential to such a piece of work, care must be taken not to offend their susceptibilities. For example, strict observation of the sabbath may be required in the Hebrides. Other unconnected difficulties may arise, as over the supply of suitable maps: in the 1950s the only available large-scale plans of some parts of the Outer Hebrides were those of the original survey carried out a hundred years previously, which failed to show contours or to give any indication of the resettlement that had sometimes followed the clearances. Finally, when all the data has been collected, they have to be suitably processed for interpretation. The development of computer programming has made this much easier than it used to be; but even so the combination of statistical, historical and spatial types of information makes analysis very complex, and helps to explain why much research never reaches general publication (Moisley, 1961, 1962a; Wheeler, 1964; Bailey and Wheeler, 1973).

(c) *The facts revealed.* The introduction to the published report (Caird, 1958) on the Park investigation points to some basic facts inhibiting previous investigations:

> Neither croft nor township boundaries are recorded on any published maps; these were obtained [by field investigation] so that land use and house sites could be plotted on 1:2,500 plans; details of crops and stock were recorded for every croft and an enumeration of population made for each household . . . Only in this way can a true picture of contemporary crofting be obtained, for official sources, such as the Census and annual Agricultural Statistics, deal mainly with counties, only occasionally with parishes, and never with townships. Yet it is the township groups which are the essence of Crofting. (p. iv)

After 72 years of concentrated government activity and subsidy, the

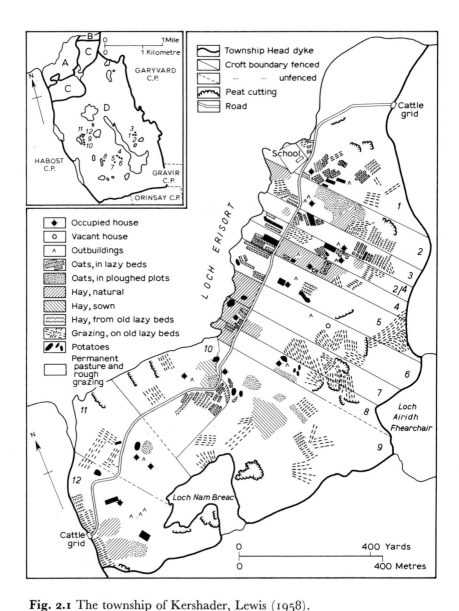

Fig. 2.1 The township of Kershader, Lewis (1958).
The main diagram shows the use of the inbye or croft land proper; lazy beds are very small strips of cultivation. The inset shows the layout of the whole township territory: A—inbye; B—common land enclosed and reseeded by the township; C—other common land fenced by the township for controlled grazing; D—residual common pastures or outrun. The former shielings or summer dwellings on the outrun are numbered to correspond with the appropriate crofts. After Caird (1958).

necessity for, and the justice of, the penultimate sentence in particular was indeed surprising.

One such township was Kershader (Fig. 2.1). Here, thirteen crofts varying in size from 1·8 to 7·3 hectares inbye, covered a total of 55·9 hectares, of which 9·9 hectares were in tillage, 3·8 hectares were under sown grass and a further 8·7 hectares of natural grass were hayed. To this was appended 792·8 hectares of common grazing, a little of which, as Fig. 2.1 shows, had been enclosed and improved. In fact, there were eleven functioning agricultural units, with 123 cattle beasts and 641 adult sheep to a total population of 53 persons, of whom only four were genuine full-time farmers. An agricultural economist would probably say that the whole township really offers only sufficient resources for one viable full-time modern agricultural unit; but if that were translated into reality what would become of the remaining families, and if they were displaced what would happen to the residual farming family bereft of society? This particular dilemma is repeated all over the Highlands, and while the Highland and Islands Development Board has persuaded government and private industry to invest heavily in certain core areas or growth points, notably the zone round the inner end of the Moray Firth from Invergordon through Inverness and hopefully as far as Nairn, the Crofters Commission (1963) has declared that 'We have come to the conclusion that there is a wide band of territory, namely the western

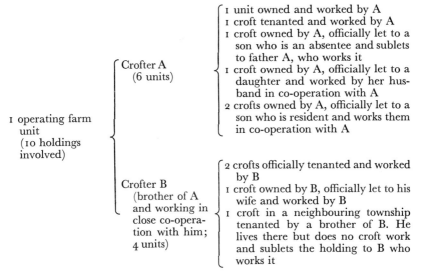

Fig. 2.2 A multiple crofting unit; all resident members of the family co-operate in operation. After Wheeler (1964).

mainland half of the Crofting Counties, in which it would be wasteful to conduct further drives to promote land improvement or re-organization' (pp. 24–5).

Since most crofts are too small to form viable farming units under modern conditions, such reorganization would normally have to take the form of amalgamation and enlargement of holdings. The ideal process, favoured in official measures, would be to amalgamate neighbouring crofts, to re-form farm boundaries, and, where necessary, to re-allocate land, so as to produce convenient, consolidated units (Osborne and Wheeler, 1969; Waites, Wheeler and Giggs, 1971). But unofficial reaction to modern economic changes within the Highlands has been to establish links between crofts that derive more often from family relationships or individual friendships, as is exemplified in Fig. 2.2. Such amalgamation is often hidden from official eyes (it was, indeed, illegal in the Crofting Counties until 1961), and certainly cannot be explained except through such painstaking work as that of the Crofting Survey. Yet these details of agricultural organization are essential to a proper understanding of the existing system, and therefore to the formulation of plans for reform which have at least some chance of succeeding both socially and economically.

It must be admitted regretfully that the value of such geographical research has tended all too often to be appreciated more by the administrations of developing lands than of Great Britain, though it is true that geographers have been drawn into land-use planning research at various levels of recent years even in this country (Gaskin, 1969). The following case, however, exemplifies what happens when the agricultural geographer turns his attention to the developing lands.

3 Case study two—Nigeria

(a) *The official data base.* Nigeria is not the largest African country in area, but with an estimated 66,000,000 inhabitants in 1970 it has by far the largest population. Over three-quarters of these are engaged in agriculture; their activities utilize three-quarters of the national land surface, and produce two-thirds of the national income. Agriculture, therefore, is certainly a legitimate subject of enquiry for geographers interested in applying their knowledge to the problems of development in such a country, and they might well begin by looking for official statistics. They would then find that the first countrywide information dealing with such matters as the number of people engaged in agriculture, the area of cultivated and uncultivated land, and livestock numbers, was collected in connection with the 1931 census of population, but that the results were published only for large administrative units, and that many data were in fact estimates. The first national agricultural census proper was that of 1950–51, but again enquiry reveals that it was avowedly only a sample census, covering about one per cent of Nigerian villages, and that it was 'largely supervised by local agricultural and administrative officers working with untrained but literate personnel locally available' (Agboola, 1965, p. 55). The next attempt at a national census was undertaken 1955–60, but even so it covered only one and a half per cent of the villages. Since this obviously provided inadequate information for development plans, a continuing programme of rural economic surveys was undertaken with the help of the United States Agency for International Development (U.S.A.I.D.) from 1963 to 1964, when approximately five per cent of the adult male population was sampled (Coppock, 1964a; Prothero, 1955; U.N.F.A.O., 1966), to 1968, when civil war supervened.

This summary indicates that in establishing the facts in Nigeria, the main problems are sporadic occurrence of official data in time and space, varied dependability and comparability of data, and the narrow range of information collected. But if official statistics are of limited, though real, use, what other sources might the geographer attempt to exploit? Since agricultural products, especially cocoa, ground nuts and palm oil, have long formed staple Nigerian exports, he will find runs of excise figures, and for those crops dealt with through marketing boards he will find records of purchases and sales; but such information is deficient areally and is of little assistance in analysing the organizational and social aspects of agriculture. In some cases special studies have been made of the current problems of important branches of agriculture or of their historical development (Hill, 1963, 1970; Hunter 1961, 1963), and these may be very helpful, but they are of limited application and generally leave untouched the foodcrops that provide 80 per cent of all agricultural production and account for a large part of domestic trade (Hay and Smith, 1970; Hodder and Ukwu, 1969).

(b) *The need for data collection.* The geographer, therefore, may well consider collecting data himself. He will then meet a range of problems, including lack of adequate large-scale maps (and probably of aerial photographs too); frequent difficulties of access, both to districts away from the established routes and to the outlying parts of village territories; difficulties of terrain and vegetation, which, particularly in the forested areas, may make it very hard to spot the farmlands to be investigated; lack of trained personnel; and a poor response from farmers, who may have a natural resentment of investigation into their private affairs, or may fear increased tax assessment. But over-shadowing all this, the European or American geographer will find himself facing the problems of using established methods of agricultural analysis for an alien system of agriculture and land holding, where land rather than crop may rotate, where boundaries of cultivated areas vary from year to year, where even the farmstead itself (and consequently the pattern of rural tracks) may shift periodically, and where concepts of ownership and operation may differ radically from those of Western European societies. Fig. 2.3 gives an example of the changes which an African farm may undergo within a few years. Under these circumstances, the farm is indeed an elusive entity (Coppock, 1964; Moss, 1959; Oyolese, 1968; Prothero, 1954). Similarly, inter-cropping, by which various crops may be simultaneously intermingled upon one plot, and succession-cropping, by which crops of various dates of harvest may also be intermingled, may make a conventional land-use map virtually impossible. Such a map in any case would represent only a momentary crystallization of an

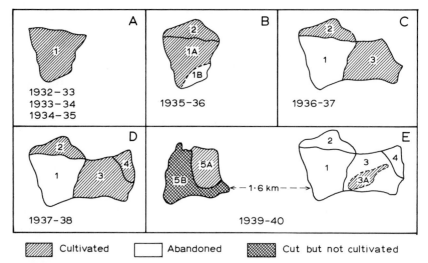

Fig. 2.3 The evolution of a farm of the Lamba tribe, Zambia, 1932–40.

1 = 2·0 ha; 1A = 1·5 ha; 1B = 0·5 ha

2 = 0·6 ha

3 = 1·9 ha; 3A = 0·6 ha

4 = 0·3 ha

5A = 1·0 ha; 5B = 1·7 ha

After Allan (1965).

immemorially fluid agricultural system not yet frozen by the pressure of growing population, of commercialization, or of close legal definition of land ownership.

III Data explanation

This recital of the difficulties encountered in the investigation of agriculture may seem discouraging, and certainly they are a handicap in the rapid accumulation of verified data, but in other ways they lend relish to research. In any case, much work by geographers and others on agriculture and its development in both advanced and developing countries does take place, and is of considerable value both as an addition to knowledge and as part of the raw materials for planning the future. None the less, it must be clear that the study of agriculture demands a very wide range of data, that such data are generally difficult to come by, and that co-operation in their collection is often required. But analysis and interpretation of data will also intimately depend upon the investigator's own assumptions and on the interpretative models he has used. Indeed, what he sees in the field may depend less on what is actually there than on what he assumes should be there for him to see. Again, the Crofting Survey offers a good example: before geographers entered the field it had never become apparent that detailed analysis of the locational and socio-spatial features of the crofting world was required just as much as was an analysis of legal and socio-economic conditions. As a result, questions which would have revealed the crucial importance of the co-operative township unit to the traditional system had not been asked. Without an appreciation of this point all attempts at alleviation and improvement were bound to be handicapped.

1 *Explanation at a broad scale*

What models, therefore, will the geographer take with him to enable him to recognize the significant features of agricultural organization when he sees them? His professional training will presumably precondition him to look for correlations between agricultural activities and spatial distributions. In many ways Chisholm's *Rural Settlement and Land Use* (1962) is still the best introduction here: 'distance . . . is the central theme around which this book is written, in an attempt to provide a systematic account of certain features of rural settlement and land use.' (p. 12). This leads Chisholm to refer to von Thünen's theory of the *isolated state* (Hall, 1966). According to this theory, concentric zones of decreasing intensity of land use would tend to develop around any isolated central settlement in a uniform plain. The proportions and types of land use would vary as transport costs grew with increasing distance from the centre. Von Thünen's detailed examples were related primarily to early nineteenth-century conditions on the North German Plain, but the general principles of his model may certainly be applied in our own time. An interesting example in a somewhat unfamiliar context may be found in Fig. 2.4a, where it is initially assumed that New South Wales is a homogeneous plain interrupted only be the sea (Rutherford, Logan and Missen, 1966), and that on this plain concentric land-use zones would form round Sydney as the central city. Fig. 2.4b shows a closer approximation to the real situation, with distortion of the ideal zones by the facts of land-type occurrence, although even then further complicating factors such as radial routes and irrigation schemes are still ignored.

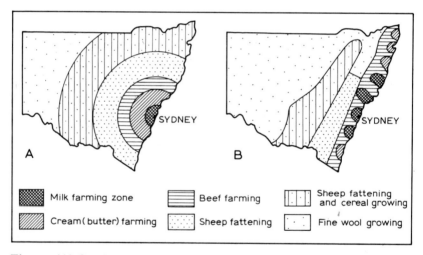

Fig. 2.4 (A) Crude and (B) refined versions of von Thünen's model of land use zonation as applied to New South Wales, Australia. After Rutherford, Logan and Missen (1966).

Von Thünen's model, as Chisholm points out, can also be applied on a smaller scale to the pattern of land exploitation round a nucleated village, where convenience of access and ease of manuring and cultivation will tend to dictate an intensive use of the land immediately round the farmsteads, whereas the most distant parts of the village territory, expensive to work in terms of time and distance, are likely to be utilized at a very low intensity. It was for this reason that at Laxton, the only remaining open-field village in England (University of Nottingham Manuscripts Department, 1969; Chambers, 1964), unimproved common grazings survived longest on land that was on the boundaries of the parish and most distant from the village, although otherwise in no way inferior to the cultivated lands closer to hand. Similar zonation can be found in many parts of the world (Prothero, 1957), as Fig. 2.5 suggests.

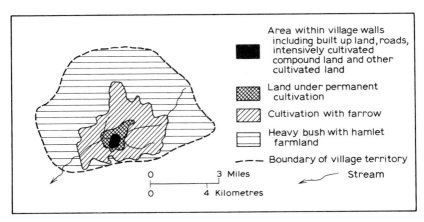

Fig. 2.5 Land-use zonation within a Nigerian village territory. After Prothero (1957).

2 *Explanation at a local scale*

The time-distance method of analysis may also be applied at the level of the individual farm. If the farmstead is regarded as 'the point of origin for all the inputs which have to be applied to the land of the farm' (Chisholm, 1962, p. 47), i.e. as the centre of organization and activity, it is reasonable to assume that a central location of the farmstead in relation to the farmland is desirable in order to minimize the time and effort (and therefore cost) of moving between farmstead and farmland. If, within a community territory such as a parish, farms are heavily fragmented, with pieces of land scattered fairly equally all over the parish, then the position of the farmstead for optimum operational efficiency under the prevailing system of land allocation will be somewhere central to that parish. Hence, a grouped settlement is likely to emerge irrespective of the more frequently cited considerations of defence, social interaction, cultural affiliation or supply of services, though these may also be relevant. It is not surprising that nucleated rural settlement is often associated with farm fragmentation, and that extreme fragmentation, especially in strip form, tends to go with a very high degree of nucleation.

Such fragmentation is normally regarded as a hindrance to modern agricultural improvement, not only because of the cumulative cost of travel, but also because small fragments, possibly difficult of access and inconvenient in shape, are difficult to incorporate in a rational, improved and mechanized agricultural system. Hence, consolidation is usually supported by advocates of rural reform, but this is hard to reconcile with nucleation if the individual farmstead is to retain even an approximately central location; a stellate layout is the only compromise, and that is feasible only for a small number of units if the individual farm is not to become inconveniently elongated. The remaining consolidated blocks must therefore lie beyond the stellate zone, as in the Danish example (Thorpe, 1951) quoted in Fig. 2.6, and if these are to be worked efficiently it is necessary for their associated farmsteads to be moved out of the nucleated settlement onto the new blocks of land. In this way, reform of farm structure is often associated with decline of the central village. In some cases, the desire of local people to continue their long-established patterns of social life which depend upon frequent face-to-face contacts in a centralized community may frustrate otherwise desirable agrarian reform, as in parts of the Italian *Mezzogiorno* (Diem, 1963; Franklin, 1969; Allan and McLennan, 1970).

This discussion assumes, of course, that the individual farm, probably the family farm, will continue to be the prevailing model. If a different model is assumed—as for example in many communist countries, where efficiency of agricultural operation is sought through the amalgamation of separate farms into very large collectives and state farms—spatial and social conditions are entirely changed. A nucleated settlement then remains desirable; from this, direction and labour allocation is easy and, with mechanized transport, workers may commute considerable distances daily in order to carry out agricultural operations on the further parts of the unit.

The relationship between consolidation and shape is complex. Consolidation can be measured by an index in which complete consolidation is represented by 100 (Simmons, 1964); but even complete consolidation may result in unsatisfactory shape, that is, in undue elongation or irregularity. A holding should therefore ideally be compact as well as consolidated, and an index of measurement has been devised (Blair and Bliss, 1967). The most compact shape is obviously a circle, with the shortest possible perimeter in relation to area, but other than in forest clearings or initial colonizing sites this shape is rarely approached in practice. The square or subrectangular

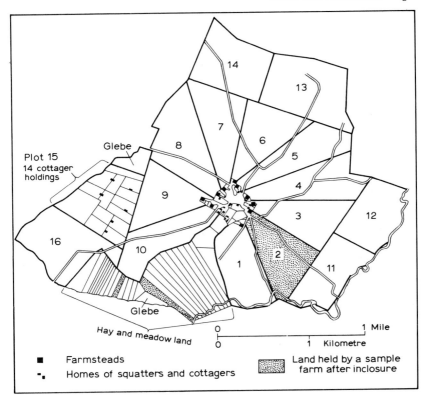

Fig. 2.6 A Danish village enclosure plan (Østerstillinge, Sjaelland, 1805). This shows the old stripped open-fields divided into consolidated blocks. There are stellate blocks round the old nucleated village, rectangular blocks on the borders of the village territory, and one group of small-holdings for those who were squatters or cottars rather than true farmers. The only survival of the strip pattern is the hay and meadow area. Farmsteads were generally relocated where necessary after 1805 on their associated block of land. After Thorpe (1951).

shape is usually the nearest practicable approach. The ideal position for the farmstead would then presumably be in the middle of such a holding, thereby reducing to a minimum time–distance costs within the farm territory. On the other hand, the farms we are accustomed to in Britain are usually divided into separate operating sub-units or fields, and Fig. 2.7 shows the effects of two simple contrasted field patterns with the farmstead in a central position (A and B) in which case B produces least cumulative distances to the field centres. However, a farmstead in such a central position requires access to the public highway by a substantial private road, thus adding to investment costs. It could alternatively have been sited directly alongside the public road, as in C and D, with lighter field tracks for internal farm access. Not surprisingly, D produces a smaller cumulative distance than C, but both C and D have greater cumulative distances than either A or B even allowing for the inclusion of the farm access road.

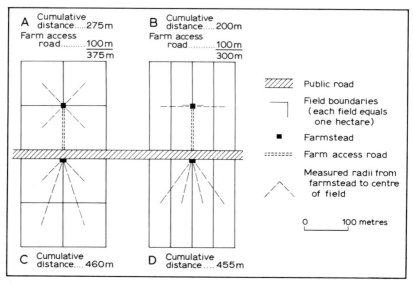

Fig. 2.7 The effects of differing field patterns upon cumulative distance measured from farmstead to the centre of each field.

Although the method of measurement is greatly simplified, these situations are not far from reality, while the principles involved are very real and govern costs, access to fields and types of land use.

3   Some further dynamics of the agrarian landscape

Of relevance to studies of farm layout there is under British planning law another rapidly growing problem that should not be forgotten by geographers: the conflict between agriculture and amenity. This arises, on the one hand, from the increasing mobility, leisure (see also Ch. 10) and education of the British public and, on the other, from changes in crop and animal husbandry, in farm layout, and in farmstead design.

Improved crop husbandry, by the use of herbicides and pesticides, and by the resulting simplification of rotations, has eliminated much interesting variety in vegetation and in animal life, and has thus encouraged a certain flatness of tone in the modern landscape. This has also been emphasized by the spread of arable associated with the growth of factory methods of stock farming: we already have battery hens and broiler chickens; intensive pig rearing and fattening has become the rule; many dairy herds are virtually continuously housed; and we appear to be on the verge of evolving battery methods of sheep farming. It is quite possible that in ten years' time livestock will rarely be seen in the open. If this does come about, permanent and ley pastures will disappear from much of the landscape. Similarly, the development of larger and more efficient machinery has demanded larger fields, thus altering farm layout, and the elimination of many hedgerows and unprofitable woodland patches, while the rising value of land has forbidden the neglect of picturesque but improvable areas.

All these changes are reflected in farmstead design. Old buildings (in Britain rarely earlier than the Agricultural Revolution) are becoming increasingly ill-adjusted to modern requirements; farmers now demand cheap, clear-span structures, built of modern materials, flexible in use and easy to alter. In fact, these do frequently come under the control of local authorities, since new agricultural buildings of 465 square metres and over fall within normal planning legislation; but even so many complaints are made about unsympathetic intrusions into traditional landscapes and the loss of attractive old buildings. The geographer clearly has a wide-open opportunity here no less than in the study of the associated changes in population distribution and employment (Green, 1971; Weller, 1967).

Geographers have in general not done much work on these topics, but there are great opportunities, once farm boundaries have been established, to calculate exact time–distance values, and to measure change. This is extremely important at a time like the present, when rapidly increasing pressure of population on the land of Britain is making us acutely aware of the decline of landscape qualities which is partly induced by agricultural change. By using such sources as enclosure and tithe commutation awards, such studies could also be projected backwards, thus adding a new dimension to our understanding of what we mean by such terms as 'the Agricultural Revolution.

IV Prediction

Once the geographer has established the facts about agriculture that are relevant to his studies, and once he has defined the working models that enable him to handle these facts, prediction might be thought to be fairly easy. After all, the farmer predicts annual crop yields on the basis of soil, anticipated weather, seed strain and husbandry, and must do so with fair success if he is to remain in business. Indeed, he will have longer periods than a year in mind when considering rotations or when contemplating alteration of the structure and layout of his farm. Surely, therefore, the geographer should be able similarly to project into the future the trends which he has by now elucidated? Unfortunately, not only is he dealing with a far more complex universe of data than is the single farmer, but he is also attempting to forecast future economic and social circumstances on an altogether wider front, since one can hardly separate agriculture from the society that both sustains and depends upon it. Furthermore, if detailed analysis involves evaluation of the disadvantages as well as of the advantages of present systems, responsible projection into the future should surely also involve evaluation of the better or worse results to be expected from alternative possible lines of activity. In other words, the geographer may well find himself moving from analysis of the past to reform of the present and improvement of the future. This is, of course, the professional field of the planner; how far geographers should be formal planners is debatable, but it is undeniable that sound planning for the future must rest on an accurate assessment of the present. Perhaps this is the final social justification for the geographical study of agriculture and the agrarian landscape.

References

AGBOOLA, S. A. (1965): The collection of agricultural statistics in Nigeria. The example of the agricultural censuses of 1950–65. *Nigerian Agricultural Journal* **2**, 55.

ALLAN, K. and M. C. MCLENNAN (1970): *Regional problems and policies in Italy and France.* London.

ALLAN, W. (1965): *The African husbandman.* London.

BAILEY, P. J. M. and P. T. WHEELER (1973): Orkney: rural and urban studies in West Mainland. *Geographical Field Group Regional Studies* [in press].

BEST, R. H. (1959): *The major land uses of Great Britain.* Wye, Kent.

BEST, R. H. and J. T. COPPOCK (1962): *The changing use of land in Britain.* London.

BLAIR, D. J. and T. H. BLISS (1967): The measurement of shape in geography. *Nottingham University, Department of Geography, Quantitative Bulletin* **11**.

BRUNSKILL, R. W. (1970): *Illustrated handbook of vernacular architecture.* London.

CAIRD, J. B. (1958): Park: a geographical study of a Lewis crofting district. *Geographical Field Group Regional Studies* **2**.

CHAMBERS, J. D. (1964): *Laxton. The last English open field village.* London.

CHISHOLM, M. (1962): *Rural settlement and land use. An essay in location*. London.

COLEMAN, A. and K. R. A. MAGGS (1968): *Land use survey handbook* (fifth edition). Isle of Thanet Geographical Association.

COPPOCK, J. T. (1956): The statistical assessment of British agriculture. *Agricultural History Review* **4**, 66–79.

(1964a): Agricultural geography in Nigeria. *Nigerian Geographical Journal* **7**, 67–90.

(1964b): *An agricultural atlas of England and Wales*. London.

(1971): *An agricultural geography of Great Britain*. London.

CROFTERS COMMISSION (1963): *Annual report for 1962*. Edinburgh.

DIEM, A. (1963): *An evaluation of land reform and reclamation in Sicily*. *Canadian Geographer* **7**, 182–91.

ERNLE, LORD (1961): *English farming past and present* (sixth edition). London.

FINANCIAL TIMES (1961): Industrial Editor, 14 June.

FOUND, W. C. (1971): *A theoretical approach to rural land-use patterns*. London.

FRANKLIN, S. H. (1969): *The European peasantry: the final phase*. London.

GASKIN, M. (1969): *North East Scotland. A survey of its development potential*. Edinburgh.

GREEN, R. J. (1971): *Country planning. The future of the rural regions*. Manchester.

GREGOR, H. F. (1970): *Geography and agriculture: themes in research*. Englewood Cliffs.

HALL, P., editor (1966): *Von Thünen's isolated state*. Translated by C. M. WARTENBERG. Oxford.

HARVEY, N. (1970): *A history of farm building in England and Wales*. Newton Abbot.

HAY, A. M. and R. H. T. SMITH (1970): *Interregional trade and money flows in Nigeria, 1964*. Ibadan.

HILL, P. (1963): *The migrant cocoa-farmers of southern Ghana*. Cambridge.

(1970): *Studies in rural capitalism in West Africa*. Cambridge.

HODDER, B. W. and U. I. UKWU (1969): *Markets in West Africa*. Ibadan.

HUNTER, J. M. (1961): Akotuakrom: a case of a devastated cocoa village in Ghana. *Transactions of the Institute of British Geographers* **29**, 161–86.

HUNTER, J. M. (1963): Cocoa migration and patterns of land ownership in the Densu valley near Suhum, Ghana. *Transactions of the Institute of British Geographers* **33**, 61–87.

INTERDEPARTMENTAL COMMITTEE ON SOCIAL AND ECONOMIC RESEARCH (1958): *Guides to official sources: No. 4—agriculture and food statistics*. London.

MINISTRY OF AGRICULTURE (1946): *National farm survey of England and Wales: a summary report*. London.

(1969): *A century of agricultural statistics. Great Britain*. London.

(1970): *Agricultural statistics 1968/1969 England and Wales*. London.

MITCHISON, R. (1962): *Agricultural Sir John: the life of Sir John Sinclair of Ulster, 1754–1835*. London.

MOISLEY, H. A. (1961): Uig: a Hebridean parish. Vol. 1, *Geographical Field Group Regional Studies* **6**.

(1962a): Uig: a Hebridean parish. Vol. 2, *Geographical Field Group Regional Studies* **7**.

(1962b): The Highlands and Islands—a crofting region? *Transactions of the Institute of British Geographers* **31**, 83–95.

MOSS, R. P. (1959): Land use mapping in tropical Africa. *Nigerian Geographical Journal* **3,** 8–17.

OSBORNE, R. H. and P. T. WHEELER (1969): Rural studies in the north east Netherlands. *Geographical Field Group Regional Studies* **14.**

OYOLESE, J. O. (1968): The mapping of land use patterns from air photographs in the forest zone of Ibadan division. *Nigerian Geographical Journal* **11,** 27–37.

PROTHERO, R. M. (1954): Some problems of land use survey in Nigeria. *Economic Geography* **30,** 60–69.

(1955): The sample census of agriculture, Nigeria 1950–1951. *Geographical Journal* **121,** 197–206.

(1957): Land use at Soba, Zaria Province, Northern Nigeria. *Economic Geography* **33,** 72–86.

RUTHERFORD, J., M. I. LOGAN and G. J. MISSEN (1966): *New viewpoints in economic geography.* Sydney.

SIMMONS, J. A. (1964): An index of farm structure, with a Nottinghamshire example. *East Midland Geographer* **3,** 255–61.

STAMP, SIR L. D. (1947): *The land of Britain: its use and misuse.* London.

STOLPER, W. (1966): *Planning without facts.* Cambridge, Mass.

SYMON, J. A. (1959): *Scottish farming past and present.* Edinburgh.

THORPE, H. (1951): The influence of inclosure on the form and pattern of rural settlement in Denmark. *Transactions of the Institute of British Geographers* **17,** 111–30.

UNITED NATIONS FOOD AND AGRICULTURAL ORGANIZATION (1959, 1970): *Yearbook of agricultural production* **12** and **24.** Rome.

(1966): *Agricultural development in Nigeria 1965–1980.* Rome.

UNIVERSITY OF NOTTINGHAM MANUSCRIPTS DEPARTMENT (1969): Laxton: life in an open field village. *Archive teaching unit no. 4.*

WAITES, B., K. S. WHEELER and J. A. GIGGS (1971): *Patterns and problems in world agriculture.* Milton.

WELLER, J. (1967): *Modern agriculture and rural planning.* London.

WHEELER, P. T. (1964): The island of Unst, Shetland. *Geographical Field Group Regional Studies* **11.**

3 Water resources

Ian Douglas

Most nations, whether in humid or arid zones, have water-resource problems high on their list of national priorities, particularly as the growth of urban areas and high living standards create demands for the supply of more water to specific locations within states. Water-resources management may be taken to be the modification of the hydrological cycle for the benefit of mankind (Fig. 3.1). Water use changes the path taken by some of the water in hydrological systems. Abstraction of water from a river for spray irrigation results in more losses by evaporation and greater infiltration into the ground than would occur if the water flowed along the river into the sea. Man thus imposes changes on the naturally varying quantities of water in the different phases of the hydrological cycle. Water management is thus part of the general field of applied physical geography—the study of changes in the physical landscape and their consequences for man.

Changes in the physical landscape operate at differing rates and scales. The shortest-term variations may be those caused by changes in the water regime of the atmosphere and soils. The study of the hydrological cycle shows that the twin elements of precipitation and runoff vary almost continuously both in time and space. The rapid increase in flow in a small river following a heavy downpour shows how quickly water may be transfered from one section of the hydrological cycle to another.

A change in any of the environmental factors which affect the relationship between precipitation and runoff alters the rate at which

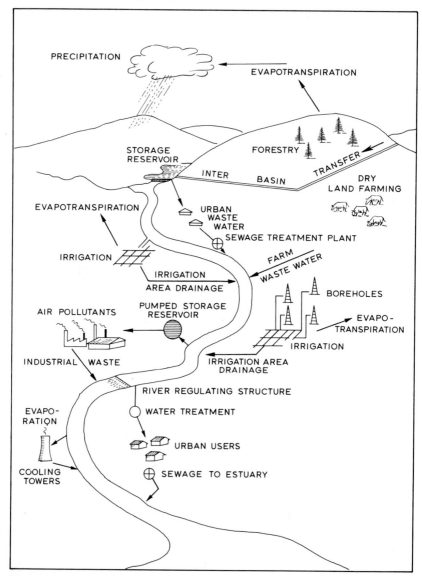

Fig. 3.1 A schematic diagram of some ways in which man alters the hydrological cycle in a river basin.

the physical landscape is changing. The extreme sensibility of the hydrological cycle to man-made changes in the landscape has made water use and conservation the subject of much geographical study. Among the most successful studies are those which view water as part of the environmental complex of potential resources available for use or misuse by man. This ecological approach to resource use sees 'the natural world as a series of interrelated systems in a state of dynamic equilibrium into which man intrudes as an unbalancing factor' (Chapman, 1969).

Two basic approaches to the study of the hydrological system have been adopted by geographers. The first concerns an evaluation of the hydrological cycle under different conditions. The second involves the study of water resources and their utilization. From these twin approaches, one largely involved with physical geography and the other more often considered to be part of economic geography, have developed studies of the interrelationships between resource use and changes in the nature of the resource.

The value of water depends on its utility as perceived by man. Water is of differing significance to people at various places and times. While water is commonly sold from carts on the Saharan fringe in Southern Tunisia, even in humid tropical West Malaysia the dry season may sometimes be so prolonged that alluvial coastal plains dry up and water is sold to villagers for five Malaysian cents a litre. Geographical studies have thus looked at not only how resources are recognized and used in different localities, but at how variations in technological awareness, social organization and political systems have affected patterns of water use through time and in different countries, and also how changes in water management affect other water users, particularly in the context of single river basins.

The following discussion will examine the different ways in which geographers have contributed to the study of water resources through investigations of the impact of resource decisions not only on the landscape but also on both groups and individuals, each utilizing water resources in different ways according to their particular resource perception (O'Riordan, 1971). This analysis is a particular illustration of the general conditions affecting the evaluation of the resources of an area (Ch. 1).

I Water-resource inventories

Although governments usually appoint civil engineers to investigate water resources, many aspects of geographical hydrology are of practical value in water-resources inventory. Pre-eminent among geographical hydrologists is Pardé, whose studies of floods and river regimes (Pardé, 1964) are internationally recognized. His close investigations of the causes and magnitude of floods, such as those in Texas (Pardé, 1962) and on the Han River in Korea (Pardé, 1968) examine the nature of meteorological events producing exceptional rainfalls, and the influence of catchment characteristics on storm precipitation : runoff ratios.

Studies of river regimes emphasize the periods of the year when water shortages are most likely to occur. The surface water-resources map of the Dawson–Fitzroy area of Queensland (Ceplecha, 1971) indicates not only the mean monthly runoff of rivers, but also the 10 and 90 percentiles of monthly runoff. Values such as these provide a good indication of the probability of drought conditions or water surplus (Fig. 3.2). Extensions of Pardé's studies for Europe by Grimm (1968), for Poland by Ziemonska (1969) and for Britain by Ward (1968)

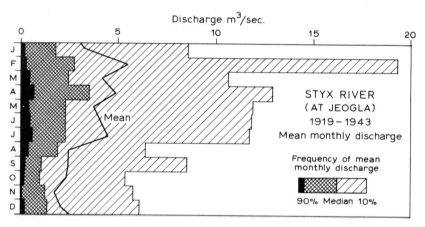

Fig. 3.2 Average discharges for each month on the Styx River at Jeogla, New South Wales, showing 10 and 90 percentiles of discharge frequency. Note the difference between the median and the 10 and 90 percentile discharges for February.

made use of the long-term records for streams gauged for seventy years or more in order to check conclusions drawn from the short records of hydrological measurements which are available for most rivers. A series of isopleth maps by Wundt (1969) show mean maximum flows of German rivers, indicating the seasons when high discharges are likely and when river flow will be sufficient to dilute industrial and sewage effluents adequately.

1 *Investigations of individual sections of the hydrological cycle*

In addition to making such statistical and cartographic analyses of hydrological data, geographers also study the components of the hydrological cycle, such as precipitation, evapotranspiration, through-flow, overland flow, soil moisture, runoff and groundwater. Although sometimes such studies are oriented towards fundamental problems in climatology or geomorphology, they all add to the store of information necessary for the wise use of water resources.

The careful examination of precipitation occurrence and intensity in Oliver's papers on the Sudan (1965a, b) shows how the value of rainfall for crops and natural plant cover varies according to the time and mode of occurrence of rain during the wet season, such as:

(i) early commencement of rains followed by partial or complete failure
(ii) late commencement of rains
(iii) extended distribution of small rainfalls throughout the wet season
(iv) well distributed but infrequent heavy rainfalls.

Each pattern of rainfall occurrence affects plant growth differently, mainly through soil-moisture variations. The time of rain is important. At night evaporation is less and more of the rain will infiltrate through the plant-root zone. As 69 per cent of the rain at Omdurman falls at night (Oliver, 1965), evaporation losses there are less than might be expected from empirical formulae which assume an equal distribution of rainfall throughout the day and night. On the other hand, Oliver also notes that the intensity of rain may sometimes be so great that runoff occurs before sufficient water has infiltrated to make up the soil-moisture deficit. Such occurrences are particularly likely in seasonally wet tropical areas where during the dry season the soil surface becomes dry, dusty, hard and cracked and the first heavy fall of rain washes all the loose material and plant debris into the cracks, thus blocking them and reducing infiltration into the ground.

The estimation of extreme rainfalls is fundamental to good water management. Any plan for a reservoir requires a knowledge of the highest expected discharges into the reservoir. As reservoirs often have to be sited in areas for which streamflow records are short, rainfalls are examined to determine the heaviest known storms. Geographers such as Lockwood (1967, 1968), have produced maps of the maximum rainfalls expected in given periods (thirty minutes up to two or three days) for various areas. Such maps are a useful guide to planners of river-basin works, providing there are sufficient data points from which to draw isopleths accurately. Two maps of rainfall intensity in Australia constructed to assist earth scientists (Jennings, 1967) are also of value when studying the development of northern Australia.

The relationship between rainfall and runoff is crucial to the development of water resources, and many geographers have participated in the study of this relationship on specific experimental catchments (see the bibliography in Ward, 1971). Such catchment research may be part of a representative basin programme, such as that organized for the International Hydrological Decade, or specially designed single-purpose projects which set out to study a section of the hydrological cycle in detail (Ward, 1971). These investigations are part of a worldwide effort to increase the store of basic hydrological data, from which may come a fuller understanding of water resources.

2   *Hydrological maps*

While catchment studies are part of an interdisciplinary effort, a more specifically geographical contribution to water-resources inventory is being made by the construction of hydrological (or hydrographic) maps (Klimaszewski, 1961; Tricart, 1963; Common and Walker, 1963). Hydrological maps aim to show the terrain characteristics which affect the movement of water, and to indicate the type of runoff which occurs on slopes and in channels. Maps at 1:1,000,000 scale showing the overall situation are supplemented by more detailed maps at 1:20,000, or 1:25,000, which portray the steepness of slopes and river channels, drainage density and stream order, vegetation, the nature of runoff on slopes, alluvial deposits and flood plains, and river channel characteristics (Tricart, 1969). Maps indicating the nature of surface runoff and lithological characteristics affecting groundwater storage help in the physical planning of rural land use by showing where changes in the vegetation are most likely to increase erosion, where stormwater runoff becomes concentrated and where possible groundwater reserves are located. With such information, soil and water conservation works may be planned to cope with extreme hydrological conditions. In the Jouch basin in Haute-Garonne, France, the various Quaternary and man-affected deposits were found to have differing hydrological characteristics and, consequently, marked contrasts in soil development (Tricart, Hirsch and Griesbach, 1966). In the oldest of the five Quaternary terraces a deeply weathered layer is overlain by a fine loam which has been easily dissected by surface runoff while the next oldest terrace has hard iron concretionary layers which inhibit vertical dissection and channel development. Each type and age of land surface causes variations in the hydrological cycle through its effects on the rainfall-runoff relationship.

Geomorphological maps (Tricart 1957; Doornkamp, 1971) and land-use survey maps (Coleman, 1970) both contribute to water-resources development by showing the locations of natural and man-made surfaces of different kinds which affect runoff. In Britain, the maps of the second land-utilization survey have been used to determine the extent of different types of vegetation (root crops, market gardens, woodland, grass, etc.) and of urban land so that estimates of potential evaporation from river catchment areas may be improved by assessing the influence of different cover types on evaporation and transpiration losses. The land-use maps, with their indications of wildscape vegetation, also help in the planning of the multi-purpose use of water resources, particularly of reservoirs and their adjacent lands.

In areas of geologically recent fluvial and marine deposition, the existence of prior stream channels may be of vital importance in the exploitation of groundwater resources (Pigram, 1971), while contrasts in grain sizes of such deposits affect their hydraulic properties and thus their suitability as groundwater aquifers. This type of terrain analysis is also essential for planning irrigation areas, as Tricart and his team of geomorphologists showed in the mid-Niger area (Tricart, 1959).

The role of geographers in water-resources inventory is twofold: participation in the interdisciplinary evaluation of the components of the hydrological cycle, and the analysis of spatial variations in the relationship between rainfall and runoff through the influence of terrain, vegetation and artificial ground cover. However, many geographical studies go beyond the realms of resources inventory into resource demand and exploitation.

II Appraisals of water supplies

The supply of water to urban areas, industrial sites, rural communities and irrigation schemes has become such a large-scale operation that national priorities have to be considered before investing in major new sources of supply. It is not surprising, therefore, that geographers have often urged the setting up of some kind of national water survey (Cunningham, 1935) which would not only make the basic water-resources inventory but would also account for the way in which water is used, transferred from one river basin to another, and eventually discharged as effluent to natural water bodies. Stressing the geo-grapher's role in putting together the supply and demand aspects of water use, Miller (1956) advocated a Water-Use Survey of Britain as a corollary of the Land-Use Survey.

Local studies of water utilization in the U.K., particularly in Nidderdale (Smith, 1966), on Teesside (Smith, 1967), in the East Riding of Yorkshire (Aylwin and Ward, 1969) and the Campsie Hills of Scotland (Cruikshank, 1965), examine the historical evolution of the present patterns of water supply and consumption, often indi-cating how many small water-supply systems develop independently, posing problems of nationalization as demand grows and larger, multi-purpose authorities are needed. For example, market gardens around large towns usually have private boreholes, thus developing a multitude of abstraction points which create difficulties for an overall water conservation scheme.

In addition to the local studies of water exploitation, several geographers have made national reviews, Balchin (1956, 1959) arguing the case for a National Water Authority in Britain in the decade prior to the creation of the Water Resources Board in 1963 to oversee the evaluation and utilization of water in Britain. Despite the new Act, problems of conflicting demands for water remain (Gregory, 1964), among the most critical being the use of reservoirs in upland areas.

1   *Multi-purpose use of water resources*

Each new proposal for reservoir construction meets with protests from a multitude of interest groups, especially from those concerned with outdoor recreation. Yet the various water supply authorities, which in Britain are still separate from the River Boards constituted under the 1963 Water Resources Act, endeavour to achieve multi-purpose use of reservoirs by allowing those recreational facilities which will minimize deterioration of water quality (Crabb and Douglas, 1970; Douglas and Crabb, 1972).

While the dense population and small area of land for recreation make the national water policy of Britain a particularly involved and crucial case, less densely settled countries have water problems of differing magnitudes. The decision to use water for irrigation in Australia has received great attention from agricultural economists (Davidson, 1969) and geographers. Schemes such as the Colorado-Big Thompson Project in the U.S.A. and the Snowy Mountains Scheme in Australia have involved massive capital outlay to turn waters through the mountains to generate hydro-electricity and irri-gate dry inland areas. Critically assessing the impact of these schemes, Loeffler (1970) comments:

> . . . both projects are already out of phase with social and economic conditions. Irrigation water is not needed to increase farm produce for which there is no market hydro-power supplies only a fraction of the total power needs and can readily be supplanted by thermal means, but the strong recreational potential of both projects has been neglected.

This critical appraisal involved consideration of alternative sources of both power and food supplied by these projects. Finding that the coal resources of Eastern Australia and the United States were adequate to meet power demands, and that overproduction of food in western countries has swamped the market to such an extent that tariff walls, such as those of the European Economic Community, reduce the export markets for the products of the irrigation areas, Loeffler demonstrated that the schemes would not have been justified under 1970 conditions. However, the planners had taken into account neither the great recreational benefits brought by the creation of reservoirs, new service roads and the opening up of new areas of country (Mercer, 1970) nor the ecological implications of such great modifications of the hydrological cycle. In the catchment areas of Australia's Snowy Mountain scheme, for example, ecosystem destruction has to be avoided by protecting the high mountain alpine flora from trampling by grazing animals, gully erosion and the effects of fire and engineering operations (Australian Academy of Science, 1957; Costin, Wimbush and Kerr, 1960; Bryant, 1969).

To achieve national use of water, surface and subsurface water management must be integrated (Pigram, 1971). In areas such as the Great Artesian Basin of Australia (Jennings, 1956) the rate of ground-water abstraction must not cause exhaustion of the subsurface reserves or deterioration of water quality. Water-quality tests have to be made before groundwater is used for irrigation or consumption by animals or men. Groundwater usually contains more dissolved mineral matter than surface waters, and may change in quality with time, or become polluted by the inflow of saline water, fertilizer wastes, septic tank seepage or mine drainage. Many geographical studies are concerned solely with water quantity, but Jennings (1956) stresses the quality of water.

2   Political and legal aspects of water-resource use

Many geographical studies rightly stress the need for rationalization of political and legal aspects of water-supply situations such as that prevailing in Northern Ireland in 1963, where there were 31 rural, 36 urban and 10 joint water supply authorities (Common, 1963). Such problems are highlighted by international or interstate rivers. In Britain there is already some political feeling over water transfers from Wales to England. In Australia the states of the Murray-Darling basin have their use of the river controlled by the joint federal and state government administered by the River Murray Commission. In America, water use in the lower Rio Grande and the Colorado River is governed by a series of agreements between Mexico and the United States which do not always cover all the ecological and social consequences of water use. Changes in channel morphology, situation and salinity downsteam of dams and the effects of irrigation, drainage, overpumping of aquifers and flooding are major problems in the international management of the Rio Grande (Day, 1970). Once water had been allocated, water rights became too inflexible to allow for the changes in water management needed to correct ecological and social consequences of earlier decisions.

Geographical water-resource appraisals thus range from the purely physical evaluation of the hydrological cycle to the study of the political, social and economic constraints on the development of water-resource management systems. By setting up analogies between water-resource situations in different places, geographers have become aware of the range of alternative paths of water-resource development (White, 1961a) and are able to compare them with the actual, narrower, practical range of choice perceived by managers. Such alternatives may be considered separately in terms of agricultural water resources, urban and industrial water supplies and flood problems.

III Investigations of agricultural water resources

As agricultural water management is at the roots of modern civiliz-ation, intricate systems of water use have evolved, such as that of the Twelve Village Irrigation System in Okayama Prefecture, Japan (Eyre, 1955). Dominating much economic, social and political life in Japanese villages are over a million irrigation co-operatives, each with its own traditional procedures and its own specific, often over-lapping, functions and administrative areas. Water guards, often members of the same family for generations, control the allocations of water, yet, as Eyre reports, discord over water distribution is a permanent feature of Twelve Village irrigation management. Any attempt to modernize water distribution, by constructing a large reservoir to increase dry-period channel flows, would disturb the social equilibrium of the 800-year-old irrigation system. As land reclamation further down river has created new irrigation areas demanding extra water, any rationalization of water distribution would deprive the Twelve Village system of some of its accumulated rights.

The Twelve Village system suggests some of the complications which face the student of water-resource use and management in the irrigated areas of Asia. Even before modernization of agricultural practices may begin, the nature of the relationship between cultivator, land and water has to be understood. This was recognized in the international project to develop the Lower Mekong Valley (White, 1963), but while most published reports deal with the engineering and economic aspects of the project, applied geographers have assessed the likely impact of changes in water management on the cultivation of the river valley. Even in a country with less entrenched rice-cultivating traditions, such as West Malaysia, the introduction of improvements in irrigation technology has to take account of the attitudes of farmers and their understanding of the way water should be distributed and used. Here provision of adequate water for dry-season cropping is not sufficient to persuade farmers to adopt double-cropping. It is equally important to convince the cultivators that traditional sowing and harvesting times, with associated festivals, have to be changed in order to achieve the higher living standards to be gained from the profits of two crops per year. In the Pahang River delta, an irrigation area remains partially as swamp because this social adjustment has not been achieved.

1 *Water-management problems in Asia*

Additional management factors considered by geographers in long-established Asian irrigation schemes include the water thefts and conflicts which vary in importance and intensity from one system to another and even within part of the same system (Vandermeer, 1971). The number and severity of water thefts and conflicts within an irrigation system increase as the adequacy of its water supply decreases, while water conflicts occur between farmers high in a system, or along a channel, and lower farmers. In a study of such spatial variations in water use in the Nan-hung gravity canal irrigation system near Pu-li City in central Taiwan, Vandermeer (1968, 1971) found that as a result of conflicts owing to thefts or to desires to steal, means of detecting thefts were incorporated in water rights and control methods used in 1930. Between 1930 and 1960 the irrigation ditch offtakes and reticulation network were altered. Opportunities for theft were reduced by the concrete lining of ditches and the provision of modern adjustable, lockable gates at offtakes from the main canal.

Improvements in agricultural technology after 1930 reduced the demand for water per unit area and enabled more fields to be planted with the first rice crop. Subsequently, the farmers formed six rotation group areas, each receiving water from several consecutive ditch headings on the canal. Each group area was further divided into team areas of approximately ten hectares which obtained water from sections of long ditches or from short ditches. Water control was carried out by teams with elected group chairmen who collaborated on operation of the rotation schemes. Farmers took turns to patrol canals at the times irrigation water was being applied, thus providing a check on illegal abstractions. Vandermeer's study concludes by showing how water thefts and conflicts are governed by complex factors, including the spatial distribution of houses and daily walking journeys to rice fields, and also by the organizational structure controlling irrigation water distribution. Significantly, where a water-flow schedule is imposed upon farmers by an outside authority without farmer participation, the co-operative spirit among farmers may be weak and thefts are more likely to occur.

Similar considerations are included in Farmer's classic study (1957) of agricultural development and related water use in the Dry Zone of Ceylon. Although Farmer recognized the slow changes in cultural practices possible in the Dry Zone, large-scale irrigation schemes were pushed ahead but not always successfully. Foster (1969) actually advocated a return to older concepts of labour-intensive irrigation schemes. These studies of Asian water-resource utilization emphasize the geographer's contribution to resource management through the recognition and assessment of the spatial linkages created when changes in resource use impinge upon the activities of social groups and on the physical environment.

2   Irrigation in Australia

Criticism of Australian irrigation policies by economists and geographers has drawn violent reaction from certain irrigation authorities. Those deeply involved with irrigation did not perceive the breadth of alternative uses of the Murray-Darling waters and did not evaluate the national costs and benefits of irrigation schemes. In Australia, many geographers feel strongly that all possible alternatives and future developments should be considered before water development policies are implemented. Langford-Smith and Rutherford (1966) insist that water use must be sufficiently flexible to be efficient in the future when economic and social conditions have changed. They point out that Murrumbidgee irrigation until now has been a costly experiment, not only to the hundreds of farmers who went bankrupt in the early years of the project, but also to the state government, which spent vast sums of money writing off capital and providing financial assistance to settlers.

In the Murrumbidgee Irrigation Area, the small, inflexible farm sizes and the lack of agricultural training proved to be severe handicaps for ex-soldier settlers born and bred in the coastal cities. By the late thirties, larger farms had replaced the original dairy-farm units and were successfully combining rice and sheep with supplementary cropping of wheat and oats, However, in March 1939 torrential storms caused serious flooding and a sudden rise of the water table which greatly affected growers of citrus, peaches, wine grapes and apricots, seriously damaging 114,000 trees and 75 hectares of vines. Further exceptional rains in 1942 and 1956 caused more damage except in areas where tile-drains had been installed. Although vegetable growing had expanded during the Second World War, some of the Italian migrants who leased small areas from Australians regarded vegetable cultivation as a way of making money to buy a larger farm. Frequent ploughing and poor soil conservation on the lighter soils of some horticultural farms caused a decline in fertility, while excessive watering raised the water table, increased salinity, and thus caused great harm to nearby fruit trees (Langford-Smith, 1966). Even though not as severe as in the South Australian portion of the Murray Basin (Crabb, 1967, 1969; Livermore, 1968, 1969), the root disease phytophtora, water table changes and salinity problems are a constant worry to the citrus grower.

Although originally the Murrumbidgee scheme suffered from over-concentration on engineering design at the expense of agricultural and socio-economic issues, the gaps between engineers, agricultural and social scientists have been gradually narrowed (Langford-Smith, 1966) and the scheme now enjoys a measure of prosperity, despite the threat to fruit-producers of the restriction of the United Kingdom market with Britain joining the European Economic Community. From Langford-Smith and Rutherford's study, it would appear that the new Coleambally Irrigation Area is benefiting from the lessons learnt on the Murrumbidgee Irrigation Area, rice-growing on large farms of 200 hectares or more being particularly successful (Ryan, 1968). Yet even so the Colleambally scheme is developing slowly and may prove to be both too inflexible, and possibly unnecessary, in view of the recession in rural industry throughout Australia. Here again, the geographer's appraisal of the physical, economic and social consequences of changing the hydrological cycle shows how major landscape changes are often initiated without adequate appraisal of resource availability and external economic factors.

IV Urban and industrial water-supply problems

Urban and industrial water needs often conflict with irrigation requirements. The best dam site for urban supplies may involve the drowning of land suitable for irrigation, or water storage for power generation may not be compatible with water control for irrigation. Thus water-resource management decisions often involve analysis of the optimum location of storage reservoirs for the combined purposes of power generation, irrigation and urban water supplies. Most storages, however, are constructed to meet a single need, save large multi-purpose schemes, such as that of the Tennessee Valley Authority. In Britain, some upland reservoirs which are basically for water supply are also used to generate small amounts of hydro-electricity, as, for example, at the Glen Finglas reservoir in Perthshire (Crabb and Douglas, 1970). New river-regulation reservoirs in Britain, such as the proposed Farndale scheme on the North York Moors, are being planned, with government encouragement to allow recreational use from the outset (Brown, 1968). Draycote Water near Rugby, Warwickshire, is an example of a modern lowland water-supply reservoir planned for boating activities.

1 *Political geography and water supplies for Singapore*

Studies of urban water supplies to Singapore illustrate the role of political geography over and above the environmental, social and economic considerations involved in all water-development projects. This Asian city-state depends for part of its water supplies on pipelines from another country with whom relations have not always been harmonious. Political events may threaten continuance of supplies, and the political boundary reduces the control the urban water-supply authority has over the management of the catchment areas. In Singapore, the original reservoirs collected streams carrying good quality, clear water out of rainforest-covered granite hills on Singapore Island (Geno-Oehlers and Wikkramatileke, 1968). These island reservoirs now form a strategic reserve of water, the Seletar reservoir having recently been greatly enlarged. Further water supplies for

Singapore are drawn from the streams of southern Johor (Fig. 3.3). The original reservoirs around the rainforest-covered Gunong Pulai are similar to those of Singapore Island and yield good quality water.

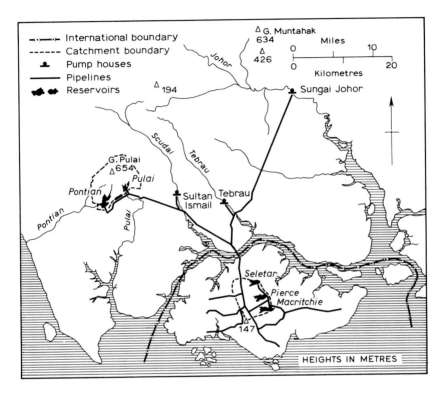

Fig. 3.3 The water-supply system of the Republic of Singapore showing reservoirs with forested catchment areas, river intake works and main international pipelines.

Most of Singapore's water is now abstracted from the Scudai, Tebrau and Johor rivers, which flow through areas of varied land use and settlement. These three rivers carry moderate to high sediment loads, with suspended sediment concentrations ranging up to 500 mg/l. (Douglas, 1967, 1970). Water is pumped from the rivers at low weirs, behind which the silt would rapidly accumulate, choking the pump intakes, save for continual cleaning of silt from the channel. Water treatment costs are thus higher than they would be if the catchment areas were entirely under control of authorities in Singapore.

2   *Industrial water-supply management*

Although domestic water consumption in cities is usually from the reticulated supply, water for industry is often drawn from private surface or groundwater sources. Not all industrial plant managers are aware of the range of alternative sources of water supply and means of conserving or recycling water. Wong (1969) examines the theoretical set of adjustments to water use industrial managers may make by recycling water, re-using water, controlling water losses, substitution or process change. Each of these forms of adjustment involves distinct technical changes in water use. Recycling involves using the water in the same process again and again through cooling systems, while the re-use of water involves the routing of water, with or without treatment, from one process to the next or to different processes for further use. Water losses may be cut by eliminating waste and suppression of evaporation. In many processes, good-quality water supplies may be replaced by stormwater, brackish or sea-water. Finally, managers may change manufacturing processes or use air, oils or other liquids to conserve water.

From a questionnaire survey of industrial water users in the Chicago area of Illinois, U.S.A., Wong (1969) found that self-supplied surface water was the largest single source of supply. Water-management decisions were dominated by economic and technological factors and hampered by a lack of knowledge of groundwater resources. Factor analysis of the questionnaire showed four basic dimensions affecting the decisions of industrial water users:

(i)   an economic-technological dimension
(ii)   a water-supply dimension
(iii) a social guide dimension
(iv)   a temporal dimension.

Industrial feasibility of alternatives in terms of cost and technical ability, as recognized in the first factor, reflects the way in which adjustments in water use generally follow awareness of the high cost of water treatment, increased cost of water supply and the need for quality control. Under these situations the adjustments perceived and adopted are usually a combination of recycling, re-use and elimination of losses. While recycling and re-use reduce pumping costs, some of the maintenance costs are saved by eliminating losses. Pollution loads in effluent are also cut by recycling and re-use, which also reduce the costs associated with treatment, labour, fuel and other overheads.

An examination of managerial attitudes helps to ascertain the possibility of introducing new sources of water or ways of improving water quality. Changing policies can change water-consumption patterns. For example, set water rates can be replaced by rates based on the volume consumed, while effluent dischargers may be fined if their waters are below quality requirements. Increased cost of effluent disposal may force the industrial manager to eliminate some of his effluent by adopting a combination of the recycling, re-use and loss elimination adjustments described earlier.

3   Water supplies for city growth

Regional development involves considering water supplies for new areas, in addition to better management in well-established localities. In Australia, with its large cities and small country towns, urban decentralization is a much-discussed objective of state governments. If inland cities are to grow, adequate water supplies are essential, such considerations having been fundamental as early as 1910, in the selection of the site for the federal capital at Canberra. As part of the discussion of the feasibility of decentralization in New South Wales, Hobbs and Woolmington (1972) have put forward a model based on the water consumption of Newcastle, 165 km north of Sydney. In 1965 gross per capita water consumption in Newcastle was 700 litres per day. On the basis of a consumption of 700 litres per day the annual gross consumption for a theoretical town would be:

$$N(365 \times 700) \text{ litres}$$

where N is the population of the theoretical town. If the ideal decentralized city has a population of 100,000, the annual water need is thus 21,920 million litres. Based on the discharge into the Chichester dam, near Newcastle, required to supply 700 litres per day per capita, a water opportunity ratio of 3.7:1 is derived, indicating that 80,640,000 m³ mean annual discharge is required to meet the annual water need of 21,920 million litres in the theoretical city of 100,000 population. By relating runoff to catchment area, Hobbs and Woolmington apply their model to different localities of New South Wales, finding that the coastal streams of New South Wales provide the maximum possible water opportunity for decentralized urban development, particularly the combined Clarence and Richmond River areas on the North Coast. However, the inland towns of Wagga Wagga and Albury, through the water-control enterprises already completed on the Murrumbidgee and Murray, also offer opportunities for decentralization. Woolmington (1971) shows that the volume of water in the Murray at Albury is sufficient to sustain a theoretical population of over 5,000,000. With a favourable marketing position, the Albury–Wodonga area is thus potentially an excellent prospect for urban growth, providing that attitudes towards the use of Murray River waters become more flexible, and that the benefits of urban water use are considered against:

(i)   existing agricultural water demands
(ii)   possible re-use of water discharged from the urban area
(iii) problems of effluent disposal
(iv)   development in the catchment area upsteam from Albury.

Such enquiries into water resources for urbanization need continual updating and modification to allow both for the increases in water consumption as living standards rise, and for the availability of new hydrological data on the variability of river flows. As Hobbs and Woolmington admit, their model does not adequately cope with the extreme variability of river discharges in New South Wales, where water storages have to hold several years' water supply to cope with successions of drought years. Nevertheless, this type of study indicates the realities of development possibilities, and could be applied to the complex water-resource problems of densely settled industrial and agricultural areas.

V Studies of floods and flood-plains

Water management involves not only the beneficial use of water resources but also the prevention, avoidance or minimization of the effects of water excess or deficiency. The value of geographical investigation of human response to natural hazards has been clearly demonstrated in the U.S.A., where flood control work has been dominated by the structural approach to flood mitigation adopted by the Corps of Engineers (White, 1961a).

The initial problem in extreme events such as flood and drought is the prediction of magnitude and frequency of such happenings. Such predictions are not always reliable. Burton and Kates (1964) quote the case of the San Carlos Reservoir on the Gila River in Arizona as an example of unsuccessful estimating. Although the reservoir was completed in 1928, it has never been filled to more than 68 per cent of its capacity and has been empty on several occasions. Though short, the thirty years of streamflow records available could have produced a sound estimate had the co-efficient of variation of streamflow been adequately examined. The overbuilding of this dam is due in part to a deficiency in developing the patterns of river-flow fluctuations. New techniques of stochastic hydrology may offer one way of approaching the problem of estimating the extreme event. Stochastic models have been widely used for studying reservoir inputs (Hufschmidt and Fiering, 1966; Beard, 1967). These statistical prediction models are not yet fully satisfactory and may be supplanted by simulation techniques based on the statistical theory of turbulence which offers an alternative approach relying on advanced probability and time series analysis (Mandelbrot and Wallis, 1968).

All too frequently, water-development and flood-mitigation projects are needed in areas with no riverflow records. One technique estimates the probable maximum precipitation by examining the heaviest storms likely to occur over the project catchments, the other technique estimates floods from the geomorphological characteristics of the drainage basin.

1   *Flood estimation from basin morphology*

The way in which runoff reaches the main stream channel depends, in part, on the nature of the drainage network. Strahler (1964) illustrates a simple case of two streams' networks with different bifurcation ratios. Stream network A in Fig. 3.4 has a bifurcation ratio of first to second order streams of 11, while stream network B has a ratio of 3. In basin B water flowing down tributaries from storm rainfall

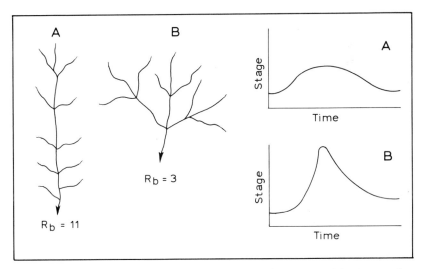

Fig. 3.4 Relationship between channel network and flood hydrographs. R_b denotes bifurcation ratio, while the graphs of stage against time show the hydrographs associated with two channel networks.

equally distributed over the catchment area would produce a storm runoff hydrograph of Type B, with a pronounced peak flow. However basin A has a drainage network of a type which allows water from the tributaries to pass down the main stream successively, so producing the flatter type of hydrograph in graph A, with a much lower peak discharge than B. This contrast in flood peak due to basin morphometry indicates the possible utility of morphometric analysis for flood estimation.

Hirsch (1962, 1965) used comparative morphometric analysis to estimate riverflows in ungauged basins, while Wong (1963) used principal component analysis of basin characteristics to show that in New England, U.S.A., as average land slope, mean altitude and tributary channel slope increased, mean annual flood increased proportionately. In a later study, Wong (1971) compared the magnitudes of floods of given recurrence intervals with stream length and average land slope, using multiple regression analysis of the form:

$$Q_T = aL^bS^c$$

where: Q_T is the T-year annual peak discharge
   L is the main stream length
   S is the average land slope
   a is the intercept
   b and c are the regression coefficients.

Using data for flows of nine different recurrence intervals ranging from 1·2 to 300 years, Wong obtained statistically significant regressions for all nine recurrence intervals. The standard error of estimate increased as the recurrence interval increased, partly because data for flows recurring at intervals greater than twenty-five years were not available for all ninety gauging stations. Once the regional relationship between basin morphometry and flood-flow is established, flood magnitudes at ungauged sites may be estimated. Nevertheless, care has to be exercised and allowance made for contrasts in local climate and, more significantly, catchment vegetation and land use. Clearing of forest from catchment areas can increase flood peaks by over 50 per cent, as water runs off the soil surface directly to the channels without being intercepted by vegetation, whereas formerly it was held in the root zone of the soil and allowed to infiltrate. For example, the large floods in the Bow River, Alberta, during the late nineteenth and early twentieth century, seem to have been at least partly induced by destruction of vegetation through fire, lumbering, and other processes (Nelson and Byrne, 1966).

2 Flood-hazard mitigation

Once the flood frequency is established, adaptation of land use to flood risk must be achieved. Many colonial settlements in the new lands of the Americas and Australasia were established in sites which appeared to settlers from Western Europe to be flood-free but which are highly flood prone. These situations of flood-plain occupance have been studied by Gilbert White and his associates at Chicago (White, 1961a, b, 1964; Burton 1962; Kates 1962, 1965; Sewell, 1965), and by geographers in New South Wales (Thorpe and Tweedie, 1956; McDonald, 1967, 1970; McDonald and Lee, 1968; Lee, 1968a, b, 1969; Davies and Lee, 1968) and New Zealand (Ericksen, 1971). Heavy downpours, lasting seven days or more, may bring well over 250 mm of rain to large areas of northern New South Wales in summer, causing widespread flooding along major rivers (Thorpe and Tweedie, 1956). The type of flooding varies with the river system, the westward flowing streams experiencing a gradual downstream migration of a flood wave of which adequate warning may be given, while the much steeper east coast streams experience rapid rises in levels and swift flood runoff of which there is little warning.

In the intensively used dairying and market-gardening areas of the flood-plains of east-coast streams, flood-plain towns such as Maitland, Kempsey and Grafton suffer considerable losses with each major flood. Geographers have participated in interdisciplinary studies of flooding in the Hunter, Macleay and Clarence Rivers (Starr, 1964). With peak flood discharges at Kempsey exceeding 10 m^3 km^2sec^{-1}, the major and medium floods of the Macleay River cannot be economically prevented, but their effects can be reduced and the almost constant influence from small nuisance floods can be prevented. To decide which type and scale of flood mitigation work should be undertaken, the benefits and costs of mitigation works were examined (McDonald, 1967). As frequent low-level flooding on the Macleay delta damages pastures only in the lower parts of the delta landscape, the benefits of mitigation may be estimated in terms of gains to rural production. As a first step, McDonald (1967) produced a map classifying the flood probability for the whole delta before and after mitigation (Fig. 3.5). The difference between the areas enclosed by the three-year recurrence isopleths before and after flood mitigation on this map shows that by preventing the small but frequent floods occurring more than once every three years, 30 per cent of the floods are avoided on some 22,000 hectares of the flood-plain, thus permitting greater agricultural production.

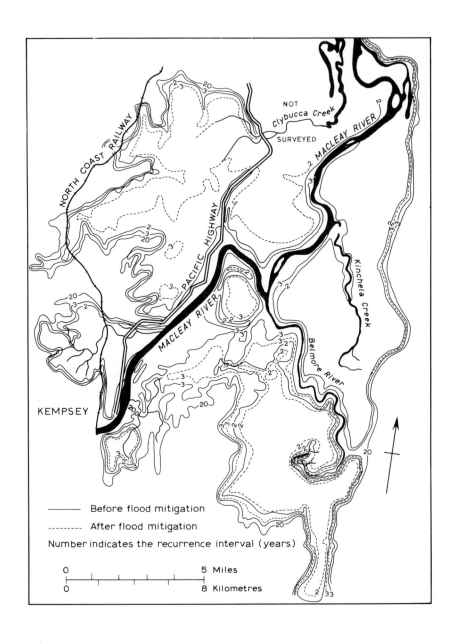

Fig. 3.5 Flood-probability classification of the Macleay River delta. After McDonald (1967).

The second phase of the Macleay investigation established the patterns of land use and the associated economic returns (McDonald and Lee, 1968; McDonald, 1970). The lands of the Macleay River flood-plain were classified into the following types:

(i)   first class arable and pasture land
(ii) first class pasture land
(iii) second class pasture land
(iv) swamp pasture.

Using the results of questionnaire surveys, the subdivision took into account the extent of maize growing and normal yields, dominant pasture types and carrying capacities, and the requirements of stock for feed of particular quality and the balance between high and low return grazing. For each type of land use, typical *enterprise budgets* comparing output and costs, to give gross margin per acre, were calculated. Flood-loss estimates involved consideration of the effects of flooding in plant growth, animal movements, milk production, harvesting costs and farm equipment.

From the hydrological data and the flood-loss information an assessment model was developed to estimate flood damage before and after the construction of embankments along the river. The model suggested that the works would reduce the damage by 62·7 per cent, saving $70,650 annually. Finally the cost/benefit ratio for the flood mitigation works was assessed, at 1 : 2 on the national level and 1 : 2·4 on the regional level. The Macleay delta study thus provided a thorough analysis of the impact of floods and flood mitigation works and showed the net benefits accruing to the community from that phase of water resource management.

3   *Alternatives in flood-plain management strategy*

A further aspect of flood-plain management involves consideration of alternative ways of adjusting to floods, either by protecting installations on the flood-plain, by adjusting activities to cope with floods, by evacuation, or by local authority zoning of the flood plain to prevent high-value structures and materials from being placed in potentially inundated areas. Most modern cities have spread onto flood-plains in the last two hundred years. London expanded onto the Thames flood-plain from the Strand down to the Embankment, so restricting the width of the river channel and making the stage (or river level) for any given discharge higher than previously. With the urbanization and intensive use of the Thames Basin above London, storm runoff reaches the river more rapidly than a hundred years ago, so that flood peaks are naturally higher. The constriction of the channel thus accentuates the high flood level, producing real flood hazard for London should a flood peak coincide with a high spring tide in the estuary. The valuable properties and public instalments bordering the Thames are thus at a high risk of flood damage, and it would be interesting to apply to London authorities and property managers the type of analysis used by White (1961b) in American cities.

White (1961b) claims that although flood problems are often investigated by collecting information on flood losses, public flood-control policies, areal extent of flood damage, nature of flooding and flood protection, such data do not explain how the present pattern of flood plain occupance has evolved and what the results and consequences of changes in government and local authority policies on flood-plain use might be. Better understanding is gained by finding out how the managers of flood-plain enterprises and occupiers of flood-plain dwellings perceive the flood risk and what attitudes they have towards methods of avoiding or minimizing flood losses. The simple case of the manager of an isolated residential property subject to occasional flooding shows how the possible alternatives to adjustment are perceived and adopted. In Table 3.1 the theoretical possibility of obtaining insurance is blocked by the general refusal of insurance companies to cover flood losses. Abatement and protection are not considered because they involve areas outside the property. Recognition of the flood hazard leads to thought of structural change, preparation for emergency action or raising the land level around the property; but because of local conviction that the government will act to mitigate floods, the property manager does not want to invest in such works. However, some structural change within the property would cut losses, by changing uses of rooms and electrical appliance connections. He is prepared to take emergency action, as he cannot bear the flood loss indefinitely, but also comes to rely on government aid in disaster times. Such a situation often occurs when people move into an unfamiliar area and buy properties without being aware of the flood risk. Often they have no other choice than to bear the loss.

This tabular scheme can be applied to the more complicated forms of flood-plain use. It attempts to describe how a manager arrives at his final decision. Reactions may differ when managers are made aware of flood risks by the publication of maps of floodable areas or when figures showing maximum flood levels are exhibited in public places. The pattern of adjustment may vary with the duration of the flood and type of enterprise. Many classes of adjustment are commonly ignored. By emphasizing the spatial linkages, White's type of analysis

Table 3.1 Elements in choice of adjustment to flood hazard by one home owner in a flood-plain (After White, 1961)

Theoretical range of adjustments	Elements in Choice							Actual choice
	Practical range of choice	Estimate of hazard	Technological trends	Economic efficiency	Spatial linkage	Others		
Bear the loss	O	−	+	−	+			R
Insurance	×							R
Land elevation	O	+ +	−	−	−			R
Structural change	O	+ +	−	+ +	+ +			R
Emergency action	O	+	−	+ +	+ +			A
Flood abatement	×							R
Flood protection	×							R
Public relief	O	+	+	+	+			A

× Blocked choice
O Open choice

+ Favourable to choice
− Unfavourable to choice

A Accepted
R Rejected

shows how changes in usage or flood conditions in one part of the flood plain affects other occupiers, thus increasing the opportunities for better management of the water resources of the flood plain.

VI Concluding review

While some geographical writing on water resources has taken a broad, descriptive view, most applied geographical investigations of water resources combine specifically geographical skills with established economic, sociological and engineering hydrology techniques to evaluate the impact of manmade changes to the hydrological cycle. Some of these geographical investigations provide new data for the planning or improvement of water-using projects, while others investigate water-resource developments in terms of national, regional, or local needs for improved production, social conditions and cultural achievement. In these evaluations the geographer is often adopting the role of critic, knowing how circumstances have changed since the inception of the scheme and being able to point out the mistakes that were made. Provided, however, that the geographer then communicates the problems and solutions found in one area to those planning water-resource projects elsewhere, he can make a real contribution to the understanding of how environment and man respond to changes in the hydrological cycle.

Applied geography thus constantly feeds information back into the general body of geographical knowledge and draws out new techniques and concepts to apply to the understanding of water resources. The constant comparison of water projects in different areas, the establishment of statistical models based on well-documented areas to help explain events in areas with little parametric data, and the examination of the links between events and objects throughout a hydrological system provide applied geography with a firm place in the interdisciplinary effort to manage the world's water resources for the benefit of mankind.

References

AUSTRALIAN ACADEMY OF SCIENCE (1957): *A Report on the condition of high mountain catchments of New South Wales*. Canberra.

AYLWIN, E. and R. C. WARD (1969): Development and utilization of water supplies in the East Riding of Yorkshire. *University of Hull, Occasional Papers in Geography* **10**.

BALCHIN, W. G. V. (1956): Water and the national economy. *The River Board's Association Year Book* **4**, 28–41.

(1959): Water supply and demand in Great Britain. *Nature* **184**, 18–20.

BEARD, L. R. (1967): Hydrologic simulation in water-yield analysis. *Proceedings Paper 5134, American Society of Civil Engineers* IR 1, 33–42.

BROWN, K. S. (1968): The amenity use of reservoirs. *Association of River Authorities Year Book, 1968*, 195–201.

BRYANT, W. G. (1969): Vegetation and ground cover trends following the exclusion of stock at three sites in the Snowy Mountains, New South Wales. *Journal of the Soil Conservation Service of New South Wales* 25, 183–98.

BURTON, I. (1962): Types of agricultural occupance of flood plains in the United States. *University of Chicago, Department of Geography, Research Paper* 75.

BURTON, I. and R. W. KATES (1964): The perception of natural hazards in resource management. *Natural Resources Journal* 3, 412–41.

CEPLECHA, V. J. (1971): The distribution of the main components of the water balance in Australia. *Australian Geographer* 9, 455–62.

CHAPMAN, J. D. (1969): Interactions between man and his resources, *in* COMMITTEE ON RESOURCES AND MAN. *Resources and Man*. San Francisco.

COLEMAN, A. (1970): Land use and water use: informal discussion. *Proceedings of the Institution of Civil Engineers* 46, 411–13.

COMMON, R. (1963): Water Resources in Northern Ireland. *Journal of the British Waterworks Association* 45, 778–99.

COMMON, R. and T. WALKER (1963): Three samples of hydrographic mapping in Northern Ireland. *Journal of the Institution of Water Engineers* 17, 395–403.

COSTIN, A. B., D. J. WIMBUSH and D. KERR (1960): Studies in catchment hydrology in the Australian Alps: II Surface runoff and soil loss. *C.S.I.R.O. Australia, Division of Plant Industry, Technical Paper* 14.

CRABB, P. (1967): Community irrigation projects in the Waikerie district of South Australia. *Geography* 52, 412–15.

(1969): Some aspects of salinity in the South Australian Upper Murray: some observations. *Australian Geographer* 11, 170–72.

CRABB, P. and I. DOUGLAS (1970): Water resources management in south-west Perthshire. *Scottish Geographical Magazine* 86, 203–8.

CRUIKSHANK, A. B. (1965): Water-resource development in the Campsies of Scotland. *Geographical Review* 55, 241–64.

CUNNINGHAM, B. (1935): National inland water survey. *Geographical Journal* 85, 531–52.

DAVIDSON, B. R. (1969): *Australia wet or dry?* Melbourne.

DAVIES, T. D. and K. W. LEE (1968): An economic appraisal of flood mitigation works to the municipality of Kempsey. *University of New England, Research Series in Applied Geography* 21.

DAY, J. C. (1970): Managing the lower Rio Grande, *University of Chicago, Department of Geography, Research Paper* 125.

DOORNKAMP, J. C. (1971): Geomorphological mapping, *in* S. H. OMINDE, editor: *Studies in East African geography and development*. London.

DOUGLAS, I. (1967): Natural and man-made erosion in the humid tropics of Australia, Malaysia and Singapore. *Publications de l'Association Internationale d'Hydrologie Scientifique* 75, 17–29.

(1970): Measurements of river erosion in West Malaysia. *Malayan Nature Journal* 23, 78–83.

DOUGLAS, I. and P. CRABB (1972): Conservation of water resources and management of catchment areas in Upland Britain. *Biological Conservation* **4**, 109–16.

ERICKSEN, N. J. (1971): Human adjustment to flood in New Zealand. *New Zealand Geographer* **27**, 105–29.

EYRE, J. D. (1955): Water controls in a Japanese irrigation system. *Geographical Review* **45**, 211–16.

FARMER, B. H. (1957): *Pioneer peasant colonization in Ceylon*. London.

FOSTER, G. J. (1969): The concept of regional development in the indigenous irrigation systems of Ceylon. *Yearbook of the Association of Pacific Coast Geographers* **31**, 91–100.

GENO-OEHLERS, J. and R. WIKKRAMATILEKE (1968): The water supply of Singapore: a fundamental resource problem. *Institute of British Geographers Special Publications* **1**, 187–202.

GREGORY, S. (1964): Water resource exploitation—policies and problems. *Geography* **49**, 310–14.

GRIMM, F. (1968): Das Abflussverhalten in Europa—Typen und regionale Gliederung. *Wissenschaftliche Veröffentlichungen des deutschen Instituts für Länderkunde Neue Folge* **25/26**, 18–180.

HIRSCH, F. (1962): Méthode de prévision des débits des cours d'eau par l'analyse morphométrique des réseaux fluviatiles. *Revue de Géomorphologie dynamique* **13**, 97–106.

(1965): Application de la morphométrie à l'hydrologie. *Revue de Géomorphologie dynamique* **15**, 172–5.

HOBBS, J. and E. WOOLMINGTON (1972): A water model for urban decentralization in New South Wales. *Annals of the Association of American Geographers* **62**, 37–41.

HUFSCHMIDT, M. M. and M. B. FIERING (1966): *Simulation techniques for design of water-resource systems*. Cambridge, Mass.

JENNINGS, J. N. (1956): Water policy for the Great Artesian Basin. *Geographical Studies* **3**, 127–32.

(1967): Two maps of rainfall intensity in Australia. *Australian Geographer* **10**, 256–62.

KATES, R. W. (1962): Hazard and choice perception in flood-plain management. *University of Chicago, Department of Geography, Research Paper* **78**.

(1965): Industrial flood losses: damage estimation in the Lehigh Valley. *University of Chicago, Department of Geography, Research Paper* **98**.

KLIMASZEWSKI, M. (1961): The problems of the geomorphological and hydrographic map on the example of the Upper Silesian industrial district. *Polish Geographical Studies* **25**, 73–81.

LANGFORD-SMITH, T. (1966): Murrumbidgee Irrigation Settlement: A study of irrigation planning, establishment, and growth, *in* T. LANGFORD-SMITH and J. RUTHERFORD: *Water and land, two case studies in irrigation*. Canberra.

LANGFORD-SMITH, T. and J. RUTHERFORD (1966): *Water and land, two case studies in irrigation*. Canberra.

LEE, K. W. (1968a): Problems of secondary drainage on the Macleay flood-plain, and their possible solution by the Macleay River County Council with extended powers. *University of New England, Research Series in Applied Geography* **22**.

(1968b): Farm level drainage on the Lower Macleay. *University of New England, Research series in Applied Geography* **23**.

(1969): River improvement works above tidal influence in the Macleay Valley. *University of New England, Research Series in Applied Geography* **24**.

LIVERMORE, J. F. (1968): Aspects of salinity in the South Australian Upper Murray. *Australian Geographer* **10**, 520–22.

(1969): Some aspects of salinity in the South Australian Upper Murray: A reply. *Australian Geographer* **11**, 173–5.

LOCKWOOD, J. G. (1967): Probable maximum 24-hour precipitation over Malaya by statistical methods. *Meteorological Magazine* **96**, 11–19.

(1968): Extreme rainfalls. *Weather* **23**, 284–9.

LOEFFLER, M. J. (1970): Australian–American interbasin water transfer. *Annals of the Association of American Geographers* **60**, 493–516.

MANDELBROT, B. and J. R. WALLIS (1968): Noah, Joseph and operational hydrology. *Water Resources Research* **4**, 909–14.

MCDONALD, G. T. (1967): A report on the hydrological implications of flood mitigation works on the flood-plain of Macleay River below Kempsey. *University of New England, Research Series in Applied Geography* **9**.

(1970): Agricultural flood damage assessment: a review and investigation of a simulation method. *Review of Marketing and Agricultural Economics* **38**, 105–20.

MCDONALD, G. T. and K. W. LEE (1968): An economic appraisal of flood mitigation works on the Macleay flood-plain. *University of New England, Research Series in Applied Geography* **10**.

MERCER, D. C. (1970): Outdoor recreation and the mountains of mainland southeastern Australia. *Geography* **55**, 78–80.

MILLER, A. A. (1956): The use and misuse of climatic resources. *Advancement of Science* **13**, 56–66.

NELSON, J. G. and A. R. BYRNE (1966): Man as an instrument of landscape change: fires, floods and national parks in the Bow Valley, Alberta. *Geographical Review* **56**, 226–38.

OLIVER, J. (1965a): Evaporation losses and rainfall regimes in Central and North Sudan. *Weather* **20**, 58–64.

(1965b): The climate of Khartoum province. *Sudan Notes and Records* **64**, 90–129.

O'RIORDAN, T. (1971): Perspectives on resource management. *Monographs in spatial and environmental systems analysis* **3**. London.

PARDÉ, M. (1962): Sur les crues exorbitantes qu'épreuvent certaines rivières des Étas-Unis, notamment du Texas. *Bulletin de l'Association Internationale d'Hydrologie Scientifique* **7**, 17–33.

(1964): *Fleuves et rivières.* Paris.

(1968): Les crues phènomènales du Han en Corée. *Annales de Géographie* **77**, 194–202.

PIGRAM, J. J. J. (1971): The adequacy of groundwater resources on the Liverpool Plains, northwestern New South Wales. *University of New England, Research Series in Applied Geography* **34**.

RYAN, J. G. (1968): Economics of the development of large area farms on the Coleambally Irrigation Area. *Division of Marketing and Agricultural Economics Miscellaneous Bulletin* **5**. Sydney.

SEWELL, W. R. D. (1965): Water management and floods in the Fraser River Basin. *University of Chicago, Department of Geography, Research Paper* **105**.

SMITH, K. (1966): Water resources of Nidderdale. *University of Liverpool, Department of Geography, Research Paper* **4**.

(1967): The availability of water on Teesside. *Journal of the British Waterworks Association* **49**, 481–9.

STARR, J. T. (1964): River Basin planning in Australia. *Geographical Review* **54**, 117–18.

STRAHLER, A. N. (1964): Quantitative geomorphology of drainage basins and channel networks *in* VEN TE CHOW, editor. *Handbook of Applied Hydrology*. New York.

THORPE, E. W. and A. D. TWEEDIE (1956): The New South Wales floods of February, 1955. *Australian Geographer* **6**, 3–13.

TRICART, J. (1957): Un nouvel instrument au service de l'ingénieur: les cartes géomorphologiques. *Le Génie Civil* **134**, 85–8, 110–12, 127–9.

(1959): Enquête sur les organismes faisant des recherches de géomorphologie appliquée. *Revue de Géomorphologie dynamique* **10**, 85–96.

(1963): Detailed hydrological mapping and its usefulness in river regime studies. *La Houille Blanche* **18**, 417–22.

(1969): Méthode de cartographie au 1/1,000,000 du contexte hydrologique élaborée au Centre de Géographie Appliquée, Université de Strasbourg, *Mélanges offerts á Maurice Pardé*, Gap, 671–82.

TRICART, J., A. R. HIRSCH and J. C. GRIESBACH (1966): La Géomorphologie de Bassin du Touch (haute-Garonne), ses implications pédologiques et hydrologiques. *Revue géographique de Pyrénées et du Sud-Ouest* **37**, 5–46.

VANDERMEER, C. (1968): Changing water control in a Taiwanese ricefield irrigation system. *Annals of the Association of American Geographers* **58**, 720–47.

(1971): Water thievery in a rice irrigation system in Taiwan. *Annals of the Association of American Geographers* **61**, 156–79.

WARD, R. C. (1968): Some runoff characteristics of British rivers. *Journal of Hydrology* **6**, 358–72.

(1971): Small watershed experiments: an appraisal of concepts and research development. *University of Hull, Occasional Papers in Geography* **18**.

WHITE, G. F. (1961a): The choice of use in resource management. *Natural Resources Journal* **1**, 23–40.

Editor (1961b): Papers on flood problems. *University of Chicago, Department of Geography, Research Paper* **70**.

(1963): Contributions of geographical analysis to river basin development. *Geographical Journal* **129**, 412–36.

(1964): Choice of adjustment to floods. *University of Chicago, Department of Geography, Research Paper* **93**.

WOOLMINGTON, E. (1971): A resource base for urbanization in the Murray River Basin. *Paper Presented at the Murray Waters Symposium*. Canberra.

WONG, S. T. (1963): A multivariate statistical model for predicting mean annual flood in New England. *Annals of the Association of American Geographers* **53**, 298–311.

(1969): Perception of choice and factors affecting industrial water supply decisions in northeastern Illinois. *University of Chicago, Department of Geography, Research Paper* **117**.

(1971): Effect of stream size and average land slope on the recurrence of floods in New England. *Geographical Analysis* **3**, 77–83.

WUNDT, W. (1969): Die mittleren Hochwasserspenden in Deutschland in Karten für das Abflussjahr und dessen Winter und Sommer Dargestelt, *in Mélanges offerts à Maurice Pardé*. Gap. 729–36.

ZIEMONSKA, J. (1969): Sur le régime des cours d'eau des Tatras occidentales polonaises, *in Mélanges offerts à Maurice Pardé*. Gap. 743-50.

4 Urban land evaluation

David Thomas

The nature and distribution of urban land has always been a centre of geographical attention. Since the early days of the subject's emergence as an academic discipline, studies of the sites and situations of towns and cities have dotted geographical literature. As time has gone by, such studies have tended to become more rigorous and certainly more focused upon the particular problems to which geographers have given high priority—though, if the truth be admitted, urban studies have also been given strong coherence by the seemingly inexplicable, but none the less marked, fashion-phases through which urban geography itself has passed.

In Britain, the investigation of urban form and growth has until quite recently proceeded mainly by empirical—that is, by descriptive-analytical—methods. It is only within the last decade or so that serious attempts have been made to theorize about urban development, and to express that work in mathematical or graphical form. Fortunately, social scientists elsewhere, and particularly those in North America, Germany, and Scandinavia, have traditionally adopted a more balanced stance in their studies, and have employed deductive-theoretical as well as inductive-empirical methods. It is upon their work that the present theoretical framework of urban geography is based, and it is to them that we must turn for an elementary understanding of the processes which determine the patterns of urban land.

I Urban land-use theory

Though the earlier urban geographers, in common with geographers involved in other studies, were concerned with the uniqueness of urban areas, those most involved with the analysis of urban land quickly realized not only that there was a perceptible and systematic order in the arrangement of land uses, but also that such ordering tended to be repeated from town to town. It was noted, for example, that the overall spatial distribution of activities within a town reflected an adjustment to the factor of distance, and that even the locational decisions of individuals—householders, factory owners, office managers (see Ch. 5 for examples), directors of retailing outlets (Ch. 6), and so on—by and large appeared to incorporate an attempt to minimize the frictional effects of travel and transfer. There was also a tendency for activities to agglomerate, some to take advantage of scale economies, others to associate for social or functional reasons with like or related land uses. In addition there was a suggestion that urban activity was organized in a way which was apparently hierarchical, and that this applied to the activity patterns within a town, as well as to the level of goods and services provided by a town. Such observations intensified the search for order within towns and cities, and it was these searches which, in turn, produced the early theories of urban land structure.

1 *Early models*

(*a*) *Zonal theory.* One of the early models of urban form arose from the observation that towns and cities developed outwards from their central areas more or less consistently in all directions. Land uses tended to sort themselves into reasonably consistent and regular concentric zones. The basis of concentric or zonal theory is the hypothesis that rents and transport costs are substitutable one for the other. Since accessibility to the central business district declines with distance outwards, then rent or land value must also decline radially. Not all activities are equally susceptible to changes in accessibility, and therefore rent curves for particular land uses will vary in slope and produce a zonal pattern akin to that shown in Fig. 4.1.

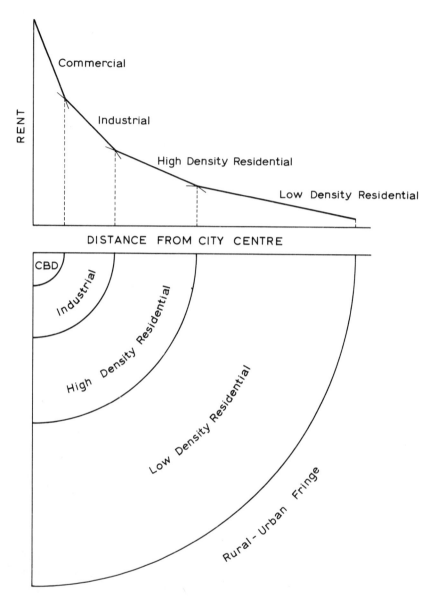

Fig. 4.1 A concentric or zonal model showing a hypothetical rent distance relationship for a city.

The best known of the zonal models was that proposed by Burgess (1925). It was based on his study of Chicago and rested also on the work of earlier writers. Burgess postulated that, within any city, land usage was so organized that zones differing in age and character occurred, always in the same order from the centre outwards. The central business district, the high-rent core of the city and the focus of commercial, social, and cultural life, contained not only the downtown retail district with large department stores and smart specialist shops, but also the main business and finance offices, civic and political headquarters, main theatres, cinemas, museums and galleries, and the most expensive hotels (Fig. 4.2A). This small and very compact area was surrounded by a transition zone of mixed land uses in which deteriorating residential property was being converted to smaller units, or, particularly on its inner margins, to other uses, such as offices or light industry. Around this zone lay a working-class residential belt containing mostly old dwellings, but with a population more stable than in the transition zone. A belt of middle-class dwellings, substantial private houses and good apartment blocks, completed the continuous built up area of the city, and was interspersed with pockets of high-cost developments. Beyond the edge of the city lay the commuters' zone, a fringe area dominated by suburban communities living in dormitory settlements. Much of the zone might still be in open country, but the villages found within it would be undergoing a change in function and becoming oriented towards the city, in which a considerable proportion of the population of the fringe-belt would be employed.

Zonal theory has come under severe criticism from later workers. A detailed review of the arguments generated by the work of Burgess has been advanced by Quinn (1940), and more recently the objections to the theory have been rehearsed by Timms (1971). Many of the criticisms have stemmed from a too literal interpretation of Burgess's formulation, a fact which Burgess himself was quick to point out, and indeed a number of authors, for example Blumenfeld (1949), have claimed to recognize a concentric patterning to the land use of a city they have studied. But it is plain that a precondition of concentric zones is a large number of closely-spaced radial routes. On simple geometric grounds, the further from the city centre, the less likely that this condition will be met, since main radial routes diverge. The tendency, observed by many zonal theory critics, of cities to assume a star-shaped form, with more rapid developments taking place along radial routes than in the interstices, is undoubtedly verifiable in the field, and it is for this reason that zonal models provide only a moderately successful description of urban areas.

(b) *Sector theory.* It was largely the unsatisfactory level of generalization provided by the zonal model which encouraged further theories based upon different assumptions. One of the most important of

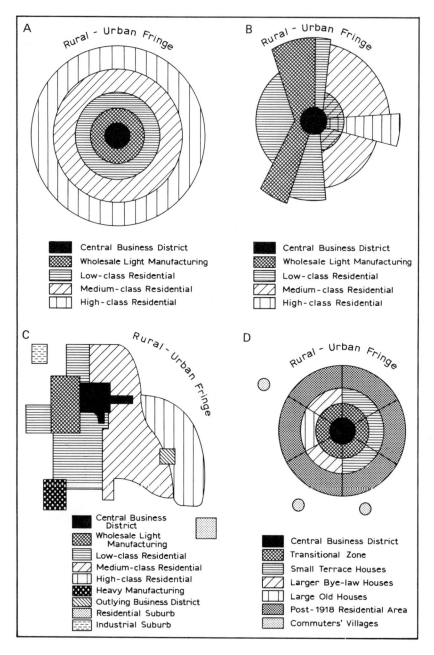

Fig. 4.2 A comparison between (A) the zonal (B) the sector and (C) the multiple-nuclei models and (D) the compromise model suggested by Mann.

these was the sector theory. All sector models are developed on the premise that the land use of a city is established by the positioning of radial routeways—an assumption quite the opposite of that adopted in the zonal model, which requires all areas of the city at equal distance from the centre to be equally accessible. Differences in accessibility—as between, say, a point on a radial route and one distant from a main route into the city centre—lead to variations in land value and hence variations in land use.

Probably the best known of the sector models was that contributed by Hoyt (1939). Unlike the zonal models, which had been strongly influenced in their development by the broad framework within which the Chicago school of ecologists operated, Hoyt's sector model was sharply focused upon rent variations within cities. He argued that similar land uses agglomerate about particular radial routes from the centre. Once contrasts arose near the city centre, these were perpetuated as the urban area expanded to form distinctive sectors. So a high-class residential district, or a light manufacturing area, would extend itself by new growth on its outer arc (Fig. 4.2B). The greater flexibility offered by mass private transportation in the form of the motor-car has to some degree removed the impetus to rigid sectoral expansion, yet the sector model still appears to provide a better approximation to reality than the zonal models. It treats industry more realistically and accommodates the processes of residential expansion, particularly of high-class residential areas, more appropriately.

As verification of this model Hoyt was able to demonstrate, in an examination of thirty American cities, how the fashionable residential districts had migrated outwards between 1900 and 1936. Jones (1960), among others, has found the distribution of residential land in a particular city to be consistent with sector theorizing.

(c) *Multiple-nuclei theory*. It may be objected that the view of the zonal and sector models is excessively naïve; that real-world cities are complex organisms which often do not conform with the assumptions on which the models are based. One of the main postulates, for instance, is that each city has a single node from which all major outward growth takes place. In practice it is observable that cities are usually multi-centred, often having many growth nuclei with varying land-use compositions. It was to accommodate this characteristic that Harris and Ullman (1945) advanced their multiple-nuclei model.

The Harris and Ullman model had a fundamentally cellular structure, focused about growth points (Fig. 4.2C). The number and position of the growth points depended upon the size of the city, its precise function, the peculiarities of its site, and upon its historical development. The larger the city the greater in number and the more specialized were its nuclei, but regardless of city size there were certain features which tended to create nuclei and to differentiate land uses. Some activities had specialized requirements which limited their locations and hence contributed to their clustering upon particular sites. Others had a tendency to agglomerate in order to reap the economies of shared services or contacts with suppliers or markets. Similarly land uses also tended to become grouped for a very different reason, namely that some activities repel others. Inconsistent land uses were thus not found in close juxtaposition, but were inclined to gather in homogeneous districts. Finally, some activities were excluded from the high-rent areas because their rent-paying ability was low. They became confined to the low-rent districts of cities. Harris and Ullman were able to identify five distinct types of district in American cities which were associated with particular nuclei: the central business district, a wholesaling and light manufacturing area, a heavy industrial zone, residential districts of various classes, and peripheral suburbs, either dormitory or industrial. The incorporation of the effects of historical and site factors into the model produced unique elements, which makes it difficult to apply the model universally. But for particular cities it does provide an indication of the way in which the urban area develops, of the precise nature of accretions at the city's edge, and of the expected developments which might take place in association with existing nuclei.

Many authors, for example Garner (1967), have insisted that the three types of model discussed above are not mutually exclusive. In most urban or even rural areas elements of all three can be identified. Haggett (1965) cites a study of urban expansion in the rural areas of southern Cambridgeshire since World War II which revealed how the growth-distance gradients varied with distance from Cambridge, with distance from a major highway, and with distance from village centres. The result suggested support for all three models. The gentle gradient from the main city was consistent with the concentric-zone model, the steeper gradient about the main road was in line with the sector model, while the steepest gradient about the outlying villages lent substance to the multiple-nuclei model. Even in a city with as simple a structure as Calgary's, Smith (1962) was able to recognize concentric zones as well as sectors. At a theoretical level the fusion of the models is not a difficult matter. Mann (1965) has attempted to illustrate the urban structure of a typical medium-sized British city by combining the concentric and sector models

(Fig. 4.2D), while Marble (in Garrison *et al.*, 1959) has proposed a model of a star-shaped city which contains ingredients derived from the concentric, sector, and multiple-nuclei theories.

2 Gradient models

Another means of illuminating urban form is presented by the study, in another dimension as it were, of urban gradients. Simply, gradient analysis is a method of describing and generalizing urban structure. It is related to zonal and sector theory, though it has a more pronounced and quantitatively rigorous empirical foundation. It is useful because any regular and logical sequences, or changes with distance from the city centre, can be identified; they can be illustrated by line-graph and represented mathematically; and they can be employed in a broadly predictive way because the relationships observed in one city often hold true for others.

(*a*) *Population densities.* Though interest in gradient analysis stretches back to the late nineteenth century, the recent widespread attention paid to urban profiles follows the work of Clark (1951). As the result of a study of the population-density gradients of 36 cities, and using statistics ranging in time from 1801 to 1950, Clark was able to argue that, regardless of the particular land-use arrangement within a city, and regardless of the date of the study, residential population density outside the central business area of a city declined as a negative exponential function of distance from the city centre (Fig. 4.3A). A logarithmic transformation of the curve produces a straight line. Two important points should be made about Clark's generalization. First, residential densities for the central areas of cities are extrapolated from the slopes derived from the outer residential areas. Since the central areas of most cities are occupied mainly by business and public buildings, the hypothetical density used in deriving the equation for the gradient is often very much higher than the observed density. Instead of the solid curve shown in Fig. 4.3A, the density profile should more resemble the profile of a volcano cone (see broken line). Secondly, the values used for densities outside the central business area are usually calculated by taking the mean of all administrative or census districts that lie at roughly equal distances from the city centre. Since metropolitan areas particularly have highly complex structures, with many subsidiary centres in addition to the main central area, it is likely that in most of the studies apparently regular population-density gradients disguise quite important differences in urban form.

More recent work has confirmed the original empirical observa-

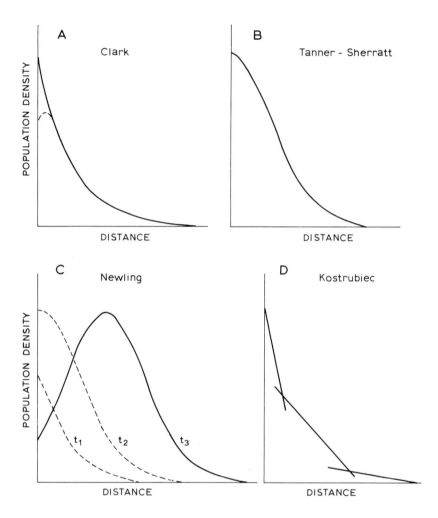

Fig. 4.3 Theoretical urban residential population density profiles as envisaged by (A) Clark (B) Tanner-Sherratt (C) Newling and (D) Kostrubiec.

tions and provided an economic rationale for declining residential densities. But, as the evidence improves, so the suggestion has arisen that Clark's negative exponential model is not a fully efficient and conceptually relevant description of urban density gradients. For example, Tanner (1961) and Sherratt (1960) have proposed that urban population densities decline exponentially as the square of distance (Fig. 4.3B), while Newling (1969) has suggested an alternative hypothesis which not only regards densities within and beyond the central business district as a continuum, but also provides a dynamic framework within which the emergence of a density crater in the central business area can be demonstrated (Fig. 4.3C, time periods t_1—t_3). Most recently Kostrubiec (1970) has outlined a three-part profile, based upon density statistics derived from London and Paris, against which the current stage of expansion of smaller urban agglomerations can be matched (Fig. 4.3D).

Most gradient analysis has been applied to western-style cities. But work by Berry, Simmons and Tennant (1963) has suggested that the population density of non-western cities may also be described in a negative exponential gradient model. Present gradients are very similar, but unlike western cities, where density gradients in the past were much steeper, the density gradients of non-western cities have remained relatively constant through time, partly due to their less flexible transport systems, which have encouraged more compact cities, partly also to the differing structure of society, which has led to a different distribution of rich and poor within a city. But it should be added that the differences noted by Berry are by no means universally agreed.

(b) *Other urban characteristics*. Clark's ideas have been considerably extended since 1951. Gradient analysis has, over the last two decades, been applied to many other urban characteristics besides residential population density, for example land value, as in the work of Berry (1963), employment densities in all or in selected industries, wholesaling, retailing, and many more. The investigations have not always yielded similar kinds of gradients and similar functions— the exponential model has appeared frequently, but is far from ubiquitous. Ajo (1964), for instance, has employed sine curves to describe population change over a distance of 130 km from central London; Korcelli (1970), on the basis of similar observations in California, has been able to propose a wave-like model of metropolitan growth; while Found (1971), in a recent contribution on urban fringe areas, has proposed (though not verified) three quite distinct distance-decay functions for speculative land value, residential land value, and the intensity of farm labour inputs.

But whatever the precise relationship, the fact remains that the various gradients, no less than the use-structure models, elucidate and provide a basis for the understanding of urban processes. They have enabled the development, in recent years, of more sophisticated dynamic models which have aimed to simulate urban growth and to predict future patterns.

3   Growth-simulation models

Most of the urban simulation models are directed to the replication and forecasting of population spatial patterns. Work has been undertaken within both deterministic and probabilistic frameworks, and a short explanatory statement needs to be made about each.

(a) *Deterministic models.* Deterministic models make strict assumptions about human behaviour. But given that these assumptions are correct, it becomes possible to predict with some degree of accuracy the future development of population in time and space. Bylund (1960), using such a model, was able to analyse the process of urbanization in the central Lappland area of Sweden and establish why, in the early phase of colonization, settlers failed to bring the best-quality land into use first, before cultivating poorer land. His hypothesis that they wished to minimize the distances between their new locations and their original settlement, between their new locations and a church, and between their new locations and a communication line, was supported by the model. Adopting these assumptions, and working from the initial settlement pattern, he was able to produce patterns which approximated closely, at each stage of development, to the actual distribution.

Though deterministic models of many different kinds have been employed with considerable success, they do have a number of disadvantages. They tend to be inflexible, they require a fairly thorough understanding of the process of change operating in any given situation, and they may also become extremely cumbersome when there are a large number of factors to be considered. The alternative, if these disadvantages are unacceptable, is the probabilistic or stochastic model.

(b) *Probabilistic models.* Probabilistic models depend upon the assumption that, though the actions of individuals may be consciously determined, the aggregate effect of human decisions, within certain constraints, may often be expressed randomly and conceived, for purposes of study, in terms of chance variations. By and large probabilistic models are more flexible and often more realistic than deterministic models, but they can also become mathematically more difficult. This problem is normally overcome by the use of Monte Carlo methods, that is, by simulating growth and change by reference to random number sequences. The random processes are often made more realistic by adjusting their operation so that they conform with empirically-observed trends.

Perhaps the best known and most ambitious practitioner of this type of modelling is Morrill. He began his work in Southern Sweden (e.g. 1963, 1965a), tracing the growth in population from 1860 to 1960. Using census material and employing principles derived from central place, industrial location, and transport location theory, he simulated population growth and predicted future distributions in twenty-year cycles with considerable success. In assigning migrants, to take one element of his model as an example, he assumed that the propensity to move from a source area was a function first of the size of the population there and secondly of its character, for instance, its age structure. The probability of migration between two areas was assumed to be closely related to their distance apart, to the difference in their attractiveness, and to their previous migration history. Given these assumptions, the chances of migration taking place from one area to another could be assessed, and the total number of migrants distributed among the destination areas by means of random number tables. In another piece of work Morrill (1965b) was able to simulate the urban growth on a city fringe. New development here, though random in direction, was visualized as being inversely related to distance from existing settlement, and inversely related to variation in land and neighbourhood quality.

Monte Carlo simulation methods are not without their difficulties. But provided that the assumptions which determine the operation of the random elements of the model are valid, and that the model is based on data that can be aggregated, they seem to provide a reasonably efficient means of representing the dynamics of urban land expansion. They are useful because they present a practical and realistic method for establishing the causal forces operating at present, because they enable prediction of future development to be made with greater precision, and because they provide a particularly effective tool for those engaged in applied studies, particularly for those, planners and others, whose task it is to make decisions about alternative courses of action to deal with current and future problems.

II Urban land appraisal

While the theoretical basis of urban geography was expanding and being refined, progress was also being made by other methods as a result of empirical studies in many towns and cities of the world. The end was the same—to understand and explain urban land patterns. It was only the means which differed. These studies were wide-ranging. When they dealt with urban land as such, they tended to focus upon the differing types of urban land and upon the particular problems which stemmed from these land uses, especially when they agglomerated into city 'quarters'. The studies also had a dynamic element in the sense that they employed description and statistical material to create in narrative form an impression of the evolving urban mass. Not only did people move from place to place, but the form and function of buildings changed through time, and with them the essential nature of cities.

1   *Urban land components*

(a) *Central area uses*. The commercial activities of the city centre have long attracted the attention of scholars, and the central business district has provided geography with what is probably its best-known set of initials (CBD). The city centre is the focus of routeways within the city and beyond. It is a highly desirable location for the many commercial, and also social and cultural, activities which depend upon accessibility. In smaller cities these many activities are intermingled, but in large urban areas there is a decided tendency for separate subdistricts to emerge devoted to government and administration, finance, retailing (cheaper shops and high-quality shops tend to group separately), and to theatres and other leisure activities.

Since most central areas developed when traffic modes were different from those of today and when traffic flows created less intense demands upon the communication arteries, the streets are relatively narrow and railway lines are often inadequate to carry present loads. Traffic congestion results and is often severe in the larger cities. To combat congestion a limited number of traffic-free precincts have been created, but a more typical response, particularly to the motor car, has been to build new car parks, often in multi-storey buildings, to construct new roads, and to introduce traffic-management schemes to facilitate traffic movement. Combined with the pressures from the central business district uses, the result has been to squeeze the less competitive uses, such as housing, out of the inner city. Cities are therefore commonly losing permanent residential population in their central areas.

(b) *Wholesaling and small-scale manufacturing uses*. In some cities intimately associated with the central business district, in others adjacent but apart from it, there occurs a zone in which wholesale activities and small-scale, workshop manufacture predominate. Both are forms of economic enterprise which require direct access to extra-city transportation, and thus a location at the focus of routeways is desirable. Both uses derive support from the city, but, unlike the true central area activities, they are not firmly drawn into the centre. Their ability to compete for city centre sites is rather lower and they therefore tend to be found on the edge of the central business district, adjacent to railway stations and major road arteries, by means of which they reach their city, regional, and national markets. The two uses contribute substantially to inner city congestion. They not only attract labour into the centre, they also generate a considerable volume of heavy commercial traffic.

Frequently, wholesaling and small manufacturing businesses are able to occupy buildings not specifically constructed for such use. They are often found in former dwellings and commonly encroach into the 'twilight' zones of older residential property surrounding the city core. These areas have not yet been subjected to the processes of urban renewal which have so modified the more desirable central locations and buildings are therefore cheap, and relatively easily obtained.

(c) *Larger-scale industrial uses.* Larger-scale industry tends to occur in wedges or nodes away from the central areas and closely associated, for supply and marketing, with rail and road arteries. The more space-extensive an industrial enterprise, the more difficult for it to be an effective competitor for city-centre locations, and the more likely that it will find a site towards the city's periphery. But since most cities have grown rapidly over the last century, once-peripheral sites are now firmly enmeshed in the suburban fabric, particularly where in the past communications, cheap sites, or strong localizing factors, such as river water for cooling or other purposes, have acted positively to create agglomerations of industrial activity. The nuisance level of larger-scale industry, in terms of noise, dirt, smell, and visual impact, have contributed markedly to the separation of this land use from others, and especially from residential land. This has assisted in producing the present clustering of industrial activity and the grouping of plant into industrial estates, sometimes speculatively built since many of the lighter industries do not require specially designed premises.

The most important dynamic element in larger-scale industry in cities is created by the observable tendency towards decentralization. The single most critical cause is space-need. Expansion of industrial output with higher living standards and growing population can only be achieved by additions to existing plant. Little spare land remains within cities, and certainly not in sufficient quantity in the right places. A continual leap-frogging into the rural-urban fringe thus results, relieving pressures within the city but transferring them to the city's edge, where industrial land impinges upon the open countryside and competes very effectively for space at the expense of agriculture, woodland, recreation space, and housing.

(*d*) *Residential uses.* Housing, certainly in terms of surface area, is the most important of city land uses. Only at the city centre is it usually superseded by other uses. Residential land is, of course, by no means homogeneous. There is the very obvious tendency for the age of houses to decline outwards. The only major exception to this trend occurs where the growing city has engulfed older and formerly quite separate settlements.

Residential land is also divisible into higher and lower-value property, each with its own characteristics. In general, higher-value houses are more likely to be on well-drained, high land and far away from industrial and other nuisances such as noise, smoke, smells and traffic. Lower-value houses are much more likely to be adjacent to factories, roads, and railway lines, and to occupy the poorer, visually less attractive areas. Perhaps the greatest contrast appears in the inner zones of residential land. Here the higher-value areas are typically undergoing partial redevelopment, sometimes with the renovation of existing property, sometimes by the replacement of existing houses with more modern, often high-rise, houses and apartments. In lower-value areas the subdivision of houses into small apartments is common. There is little redevelopment, unless the city authority itself carries it out, and generally the area deteriorates and is invaded by industrial and commercial uses seeking cheap but accessible land.

Neither is a city normally homogeneous in its social or ethnic composition. The most obvious example of this is the growth of ghettos containing minority groups. All residential areas have some measure of cohesiveness. Ghettos, extreme examples of this feature, tend to develop at the inner margins of the older low-value property, where rents are low, journeys to work are short and inexpensive, and where the living costs of low-paid workers can be reduced as much as possible. In such areas minority ethnic groups frequently cluster, but it has become evident that when economic circumstances improve the group does not break up. Ties within and pressures from without often maintain segregation, and the ghetto is contained, albeit with many differing economic groups. Alternatively there appears a tendency for ghettos to push outwards with economic advance into better and newer housing areas at greater distances from the city centre.

(e) *Minor nuclei.* Special concentrations of land uses of two kinds appear in most cities. First, in addition to the central business district, there are usually a large number of minor nodes providing a more limited range of lower order goods and services for a section of the city only. Some, in major cities, may be the equivalent in size, functions, and in tributary population of regional centres elsewhere. Others may be purely local centres providing convenience goods and services for a small trade area.

The second form of minor nucleus might be termed a 'special-use' node. Certain types of city land-use lead to the creation of identifiably different and often semi-independent communities. A major university campus within a city is a good example—almost a town within a city; the rather special type of high-class residential community which frequently develops around a city lake, or specially attractive area of open space, is another.

2 *Planning urban land*

(a) *An historical overview.* The development of distinctive areas within cities, particularly following the explosive growth of cities in the late nineteenth century and afterwards, has important consequences. Each area had its own characteristics, but each also had its own special problems. These required remedies, and gradually they came in the form of plans for the perfect city (and sometimes for the perfect society also). The motivation was a curious combination of the idealistic and the practical. On the one hand there arose in Europe, North America, and in the regions of European colonization in the southern hemisphere, a positive desire among a limited number of thinking people to create cities which were both efficient and pleasant to live in; for example, Robert Owen's proposed, but only partially executed, model settlement at New Lanark, in the Clyde valley, during the early part of the nineteenth century was followed by many other model industrial settlements later in that century, such as those at Saltaire, Yorkshire; Bournville, Birmingham; Pullman, Illinois; and Argenta Park, near Delft. On the other hand there were pressing public health problems which stemmed from the rapid and unorganized urbanization, and which had to be solved. Eventually, by the early years of the twentieth century it came to be recognized tacitly, if not always explicitly in many countries, that to achieve these economic, social and aesthetic aims, it was necessary to interfere with the market mechanisms, and also with the property rights of individuals.

No better example can be found of the gradual acceptance of statutory control over urban land usage than the growth of town planning in England and Wales. Town planning here grew out of two pressures. The first came from the protagonists of garden cities, who demanded not only new, model settlements, but also higher standards of amenity in existing towns and cities, and the enlightened development of new suburbs. The second pressure arose from the real need to meet the health and housing difficulties which had been created by the Industrial Revolution (Ashworth, 1954). In this respect, town planning in England and Wales may be regarded, certainly in its early years, as an extension of public health and housing policy.

The first piece of legislation to incorporate the term 'town planning' was enacted in 1909 (The Housing, Town Planning, etc. Act). It gave local district authorities modest permissive powers to control health standards in suburbs about to be built by allowing the introduction of by-laws more rigorous than those allowed under existing public health legislation. The local authority could also, if it wished (and few did at that time), adopt and administer planning schemes for urban areas. Pressure continued and subsequent legislation modified and eventually greatly extended this first statute. Under the Housing, Town Planning, etc., Act of 1919 the preparation of planning schemes become compulsory for some of the larger authorities, and a limited amount of joint action between authorities was permitted. An Act of 1929 allowed counties to join with the district authorities in joint schemes, while the Town and Country Planning Act of 1932 enabled planning schemes to be applied to all land, irrespective of whether it was likely to be developed or re-developed.

Though substantial progress had been made on paper, in practice there had been slower advance in the control over urban land and in the solution of its problems. It has been estimated that by 1942, while 72 per cent of England and 36 per cent of Wales was subject to interim development control, only about five per cent of England and one per cent of Wales were actually affected by operative schemes (Cullingworth, 1964). Several major towns and cities were without even an outline plan for present control and future development. But in 1947, following numerous wartime committees and commissions, a new Act was passed which transformed the situation by setting up a comprehensive and obligatory planning code. Planning was placed in the hands of the counties, each of which was required to produce a development plan. Control of development (with few exceptions aside from agriculture and forestry) was exercised by the counties, and consistently with the plan. Some parts of this Act have since been modified, principally the sections on compensation payments and betterment charges (that is, the charge levied where the value of land increases as a result of a planning decision) which were abandoned in the early 1950s; but broadly the Act, as codified in subsequent legislation, still forms the basis of planning in England and Wales.

One further enactment provided an important legislative control over urban development, namely the New Towns Act of 1946 (later consolidated in 1965). It drew its inspiration from the garden city movement and was a further practical contribution towards an integrated and balanced urban society. New towns represent a reaction against the over-crowded conditions in industrial and commerical settlements, and against the over-long work and plea-sure journeys which these urban forms sometimes impose upon their population. Hence new towns were, ideally, self-contained communi-ties set in the countryside, where housing, jobs, and all necessary facilities were provided. While new towns and their derivatives—expanded towns—may not always have achieved their highest aims, they have certainly made considerable progress in minimizing urban land-use conflicts, and consequently in creating better, happier, and more efficient lives for their inhabitants.

The foregoing historical snapshot has referred only to England and Wales. Here, control over the mechanisms of urban change began early and developed far. But elsewhere, and particularly where major urban areas have caused problems for national and local administrators, similar, though certainly not identical, means have grown up for planning towns and cities. Broadly analogous stories could thus be told for many other developed countries.

(b) *The contribution of geographers.* So far, this chapter has dealt with three closely related aspects of urban land evaluation. In the first part an attempt was made to demonstrate how urban land-use theory and urban modelling contributed to an understanding of the processes and mechanisms of urban form. It showed, among other things, how theory could assist those, like physical planners, who are concerned with controlling future events, and whose only guide is experience of the past. In the second part of the chapter attention was focused upon the empirical realities of urban areas. This revealed the real-world problems that arise in association with different parts of cities, which tend to acquire particular functions. The attempts to solve urban problems by developing legally-based planning systems has occupied the immediately foregoing section of the chapter. But it is clear that if applied geography is to be regarded as a reality in this branch of the subject, then a prerequisite is that the conclusions of theory, and the facts and tendencies observed empirically, must be introduced into the planning process by applied geographers. A comment on this fourth and important aspect completes the chapter.

Though applied urban geographers exist in all countries where geography is practised, it is convenient once more to take England and Wales as an example, not least because of the recent valuable account of the contributions of geography to physical planning by Willatts (1971). In Willatts' view, the interwar period in England and Wales was characterized by weak planning and comparatively little interest by geographers. It was only when the serious disparities in growth and prosperity between the older industrial areas and the rapidly expanding cities of the midlands and south appeared in the late 1930s that the attention of geographers was attracted. Thereafter Willatts sees three main phases of developing interaction between geographers and planning, each spanning approximately a decade.

During the forties the setting up of a comprehensive system of statutory planning required skilled manpower of many levels. Geographers were attracted in great numbers. New ministries required staff who could handle and analyse areal data, and so, after 1947, did the new or greatly expanded planning departments of local authorities. National commissions and committees of enquiry needed experts to give advice on a wide variety of topics, many with spatial aspects, and numerous statutory bodies appointed geographers to their number. The fifties, by comparison, were years of consolidation. The widespread expansion of geographical interest which characterized the previous decade could not continue. Instead there developed an interest in refining method and an increasing appreciation of the complexities of planning problems, particularly those which centred upon urban–rural relations. During the sixties geographers greatly contributed towards the new dynamic approach which grew up within planning. More geographers than ever before entered the planning profession, while those applied geographers who remained outside introduced more sophisticated means of data handling and, through theory and other means, a deeper under-standing of the processes of urban and rural land-use patterns than had previously been current. Willatts concludes that geography and planning together have made immense advances over the last three decades which can be measured not merely by the increases in the numbers of their practitioners, but by their achievements and by the techniques they have developed to handle the problems created or intensified by our modern society. It has to be confessed that Willatts is a geographer turned planner. But whether or not his judgement is accepted, his command of the facts cannot be questioned. In the field of urban land evaluation, as in many other branches of human geography, the relevance of geographical work to social and economic problems is not only an academic truth, but a practical one also. This is one of the subject's greatest strengths and one of the most hopeful auguries for its future.

References

AJO, R. (1964): London's field response in terms of population change; London's field response II. *Acta Geographica* **18**, (**2**), 1–19, (**3**), 1–25.

ASHWORTH, W. (1954): *The genesis of modern British town planning.* London.

BERRY, B. J. L. (1963): Commercial structure and commercial blight. *University of Chicago, Department of Geography, Research Paper* **85**.

BERRY, B. J. L., J. W. SIMMONS and R. J. TENNANT (1963): Urban population densities, structure and change. *Geographical Review* **53**, 389–405.

BLUMENFELD, H. (1949): On the concentric circle theory of urban growth. *Land Economics* **25**, 209–12.

BURGESS, E. W. (1925): The growth of the city, *in* R. E. PARK, E. W. BURGESS and R. A. MACKENZIE, editors: *The city.* Chicago.

BYLUND, E. (1960): Theoretical considerations regarding the distribution of settlement in inner north Sweden. *Geografiska Annaler* **42**, 225–31.

CLARK, C. (1951): Urban population densities. *Journal of the Royal Statistical Society* **Series A**, **114**, 490–96.

CULLINGWORTH, J. B. (1964): *Town and country planning in England and Wales.* London.

FOUND, W. (1971): *A theoretical approach to rural land-use patterns.* London.

GARNER, B. J. (1967): Models of urban geography and settlement, *in* R. J. CHORLEY and P. HAGGETT, editors: *Models in Geography.* London.

GARRISON, W., B. J. L. BERRY, D. F. MARBLE, J. D. NYSTUEN and R. L. MORRILL (1959): *Studies of highway development and geographic change.* Seattle.

HAGGETT, P. (1965): *Locational analysis in human geography.* London.

HARRIS, C. D. and E. L. ULLMAN (1945): The nature of cities. *Annals of the American Academy of Political Science* **242**, 7–17.

HOYT, H. (1939): *The structure and growth of residential neighbourhoods in American cities.* Washington.

JONES, E. (1960): *A social geography of Belfast.* London.

KORCELLI, P. (1970): A wave-like model of metropolitan spatial growth. *Papers, Regional Science Association* **24**, 127–38.

KOSTRUBIEC, B. (1970): Bandania roswoju przestrzennego aglomeracji miejckiej metoda profilów. *Przeglad Geograficzny* **42**, 235–48.

MANN, P. (1965): *An approach to urban sociology.* London.

MORRILL, R. L. (1963): The development and spatial distribution of towns in Sweden: an historical-predictive approach. *Annals of the Association of American Geographers* **53,** 1–14.
(1965a): *Migration and the spread and growth of urban settlement.* Lund.
(1965b): Expansion of the urban fringe: a simulation experiment. *Papers, Regional Science Association,* **15,** 185–202.

NEWLING, B. E. (1969): The spatial variation of urban population densities. *Geographical Review* **59,** 242–52.

QUINN, J. A. (1940): The Burgess zonal hypothesis and its critics *American Sociological Review* **5,** 210–18.

SHERRATT, G. G. (1960): A model for general urban growth, *in* C. W. CHURCHMAN and M. VERHULST, editors: *Management sciences, models and techniques.* New York.

SMITH, P. J. (1962): Calgary: a study in urban pattern. *Economic Geography* **38,** 315–29.

TANNER, J. C. (1961): *Factors affecting the amount of travel.* Department of Scientific and Industrial Research, London.

TIMMS, D. W. G. (1971): *The urban mosaic: towards a theory of residential differentiation.* Cambridge.

WILLATTS, E. C. (1971): Planning and geography in the last three decades. *Geographical Journal* **137,** 311–38.

5 Manpower planning

David G. Rankin

I Labour resources

Labour is one of a country's most important resources, and its effective management is essential in maintaining economic growth. For this reason, this chapter will develop a theme connected with manpower resources for manufacturing industries and will focus specifically upon the application of geographical concepts to the understanding of the operation of local and sub-regional labour markets. The essay will show how geographical methods can be of use to national and local government as well as the entrepreneur in policy formation directed at manpower planning. In addition the holistic viewpoint adopted by the geographer can contribute to the planner's task of bringing about the 'continuous and permanent improvement of the physical environment as a whole' (Brenikov, 1967, p. 485).

1   *The regional problem in the U.K.*

Since the 1930s economic growth, full employment and social equality have been the basis of economic and social policies in many countries (Kuklinski, 1970), and industrial development is one of the essential avenues through which the planned use of labour and investment has been established. The importance of industry as a promoter of economic growth has been examined on the macroscale by Myrdal (1957) and Hirschman (1958), who argue that in a free market the operation of economic forces will tend to accentuate any regional imbalance which may be present. The developing regions will present a favourable economic environment which will be attractive to investment decisions and consequently serve to increase inter-regional differentials in terms of such indices as growth in output and per capita income. If these trends towards inequality are not to develop, then government intervention is required in the form of a regional development policy.

In the United Kingdom growing concern with such problems was first translated into policy with the Special Areas Act in 1934, and a detailed appraisal on the distribution of the industrial population with specific reference to congestion in the London area was presented in 1940 (Barlow Report, 1940). A comprehensive study of regional policy had been carried out by McCrone (1969). In postwar Britain 'the relief of unemployment has been the prime motivation for these policies and areal differences in unemployment rates have been the key to regional development' (Salt, 1969, p. 93). Selective development to alleviate unemployment was the thinking basic to legislation such as the 1954 Distribution of Industries Act, the 1960 Local Employment Act and the 1966 Industrial Development Act.

To view regional inequalities in employment levels as the outcome of an imbalance between the location of population and job opportunities, and to consider that the remedy can be found in promoting either population migration or industrial movement, is a simplistic notion which in many ways evades the fundamental problem. The policies implemented so far have been directed at the control of industrial expansion and relocation through the selective issue of Board of Trade Development Certificates and the offer of financial inducements to firms which locate in the Development Areas.

The policy of taking work to the workers may not be efficient in promoting national economic growth but could be a desirable method of implementing social policy. There is little to suggest that disadvantages would be inherited by manufacturing industries through relocation (Needleman, 1965). Richardson and West (1964) favour the dual strategy of encouraging industrial relocation into the Development Areas while at the same time attempting to increase labour mobility into the expanding regions to counteract the labour shortage within them.

The major development in the U.K. during the postwar period has been the migration of population into the South-East/Midlands area, accompanied by the development of new technological industries and tertiary occupations in general. Industrial research activity has also been located within this zone. Buswell and Lewis (1970) found that 71 per cent of the private research establishments which they sampled had chosen their location 'in order to be in close proximity to the company's manufacturing plant or head office' (p. 303). The attraction of this zone can be explained by the high level of accessibility to labour (especially skilled labour) and markets, as well as the advantages of external economies gained from agglomeration and the minimization of risk and uncertainty in expansion policies. Alternatively Keeble (1970), looking at industries which have moved out of this zone, suggests that the 'spatial variation in migration to the peripheral regions can be largely explained by variations in labour availability and distance from major origin regions, while relative proximity to regional and national markets appears also to have exerted some influence upon the migration pattern' (p. 397). Labour availability is seen to play an important role in industrial location decisions in both the megalopolitan and peripheral areas; and as companies expand through acquisition and mergers then the corporate enterprises increasingly operate in terms of a world economy and regional markets diminish in significance.

The dominance of the megalopolitan belt for population concentrations is seen to be a continuing process, although within it there has been a dispersal of population to the outer suburban areas and in some cases a significant outward migration of industry (Tulpule, 1969). The core area is still the zone of greatest economic potential, but the rate of increase of manufacturing employment has been lower in the postwar period than that of the zone adjacent to it (Clark, 1966).

II Labour supply as a location factor

1   *The sub-region*

Labour availability is an important factor in determining the location of industrial establishments, especially branch plants (Luttrell, 1962). At the sub-regional level Britton (1967), in his examination of the manufacturing sector of the Bristol region's economy, found that an adequate supply of labour was the prime factor which emerged when factories were asked to rank a series of location factors in order of importance. Scope for plant expansion on the site came second, whilst advantageous freight costs, both for the input of raw materials and the output of products to the market, were ranked low in the list.

Smith (1969), studying North-West England, emphasizes the importance of labour and cites the Ford and Vauxhall motor factories, which appear to have recruited labour from other firms in the area by means of higher wages and better fringe benefits. Consequently this has not improved the unemployment situation and has left other firms with a shortage of skilled labour. Smith (1969) also points out that for some firms 'labour costs are such an important factor in their total costs that local reserves of suitable labour hidden or suggested by the level of unemployment, provide the dominant influence of locational choice' (p. 135).

An earlier study by House and Fullerton (1960) in North-East England found labour to be either an important location factor or a location problem for a wide range of industries. All firms interviewed in the electrical engineering and electrical goods industries were branch plants of firms established in other parts of the country who had required labour and sites in order to expand. The post-war location in the area of firms in the textile, leather and clothing industry had been mainly through the attraction of female labour available in the area, whereas in the metal manufacturing industry 'the problem of labour supply, especially the supply of skilled labour, constitutes perhaps the greatest single difficulty facing the iron and steel firms in the region' (House and Fullerton, 1960, p. 206) and had produced intensive inter-firm competition.

Problems of labour availability are also manifest in the rural environment, and R. Thomas (1966) discovered that the growth of some firms in rural central Wales was impeded by labour shortages. Furthermore, the inadequate public transport system precluded the exploitation of potential female labour reserves in some of the more remote villages. The response to this situation by some of the firms had been either to open branch plants in other localities or to transfer work to their main factory.

2   *Interaction among sub-regions*

The utilization of manpower resources presents problems at both the inter- and intra-regional levels, and therefore any analysis must be designed at a scale appropriate to the problem in question. Changes within the urban-industrial system can be expected to take place in three ways, 'nationally outwards from the manufacturing belt, down through the urban hierarchy and outwards from the urban centre into its surrounding urban field' (Berry and Neils, 1969, p. 295).

It has been suggested that the geographer's most important contribution can come from an understanding, not only of the spatial arrangements of phenomena, but also from the connections or linkages between them (Bunge, 1962). A knowledge of the linkages which occur between land uses helps to show how an urban area functions (Hemmens, 1966), and moreover the spatial interaction between land-use activities can be examined in terms of aggregate or individual components (Chapin, 1965). Generalizations (concepts) based on both grouped and individual data would appear to have an advantage over generalizations based on either one or the other (Allardt, 1969). Geographers are increasingly directing their attention to the behaviour of both the individual and the firm to express the relationship of activities within the city or city region.

(*a*) *Inter-firm linkages.* In the Weberian theory of location, transport costs are considered the decisive factor in determining least-cost locations. External economies of agglomeration and cheap labour cost inputs are incorporated as factors which distort the transport-cost oriented locational framework. During the 1960s geographers showed an interest in the linkage flows of both materials and information among industries, firms and plants within multi-corporate firms. Wood (1969) has developed a tentative typology of inter-firm linkage structure which Smith (1970) suggested could be represented as input costs and incorporated into a modified Weberian locational scheme.

Taylor (1970) argued that inter-firm linkages may contribute 'as a necessary set of conditions for the survival and further development of firms in their areas of birth' and not 'as forces pulling entrepreneurs to particular optimum locations' (p. 54). Townroe (1970) complements this approach by viewing linkage structure as an essential ingredient in the formulation of industrial location policy. Industries would then be evaluated in terms of the economic feasibility and desirability of movement. Studies of inter-firm linkages as elements in the promotion of regional industrial growth both in developed and developing regions is one way in which the geographer can make a valuable contribution to regional management policies, especially as this affects the efficient deployment of labour.

(*b*) *Communications and urban development.* The complex pattern of relationships in local labour markets between firms and workers results from the wide choice of residence and workplace made possible by the proliferation of both in industrialized urban areas (Dyckman, 1963). Urban expansion and the creation of extensive suburban areas, aided from the 1920s onward by a lowering of residential densities in terms of dwellings per acre, has progressed hand in hand with the increase in mobility of the population. But because of the space-consuming demands of the automobile and the facility for movement afforded, the growth has tended to an amorphous urban sprawl.

The growth of communications means that not only is the urban development process accelerated but also that urban areas not contiguous to the central city can participate more easily in central city activities, especially employment opportunities. Therefore an increase occurs in the functional interdependence between urban areas which is given focus and direction by developments originating within the central city.

(*c*) *Local industrial decentralization.* There is evidence to suggest that the local decentralization of industry, over short distances, can lead to labour supply problems (Harrison, 1967) aggravated by the loss of key workers and the expense incurred in training certain occupational categories (Keeble, 1968). Therefore the relationship between workplace and residence is of prime importance when viewed in the context of aggregate changes in urban land-use patterns and population distribution. It is the category of strategic planning which is of particular concern, because strategic plans provide the guidelines for changes in the land-use pattern based on considerations of existing and projected changes in and demands of the urban community as a whole.

III Local labour markets

The contribution of the geographer to the question of workplace-residence location and its relationship with labour supply has been either examined on an aggregate level of movement between local authority areas (Humphreys, 1965; Lawton, 1968) or has been implicit within studies concerned with the delimitation of city regions and urban hinterlands (Green, 1953; Walker, 1967).

The distribution of the population which constitutes both the labour supply and consumer market is central to economic geography (Stewart and Warntz, 1958), and one of the characteristics of the local labour force that influences production costs 'independently of the skill of individuals is the size of the group within commuting distance' (Hoover, 1963). Moreover, with an increase in establishment size, the 'significance of the local labour pool becomes magnified' (Pred, 1966, p. 79). Large establishments are absorbing a continuously greater proportion of the national industrial labour force. In the U.K. in 1935 firms employing 1,000 or more employees accounted for 21·5 per cent of the total employment in firms, but by 1961 this figure had risen to 34·5 per cent (Chisholm, 1966).

It has been recognized that the development of both public and private transportation facilities has increased the local mobility of labour and extended city catchment areas (Estall and Buchanan, 1966), but labour mobility is seldom discussed below the inter-regional level. Therefore, an attempt must be made to penetrate beneath the aggregate pattern of local journey-to-work movement to bring about a more explicit comprehension of both the spatial and non-spatial processes governing local labour mobility and in so doing improve the perception of the transactions taking place within labour markets.

Few attempts have been made to investigate labour supply at the level of the individual plant or residential community (Taaffe, Garner and Yeats, 1963; House and Knight, 1966; Grime and Starkie, 1968). A detailed penetration of labour supply areas can only be obtained from specific case studies set within a comprehensive appraisal of aggregate distributions and movement patterns. Journey-to-work activity patterns as linkages between industrial and residential land uses can provide a focus for such investigations. The situation, however, is not static because of the changes over periods of time in labour mobility.

1   *Components of the labour market*

(a) *Labour mobility.* Labour mobility is a term which encompasses occupational, industrial, social and geographical mobility which together determine an individual's propensity to move, whereas movement observed as a change in workplace or residential location is a propensity to move which is fulfilled by the presence of opportunity (Reynolds, 1951).

Because of the multivariate composition of labour mobility, it is equally essential to define its constituent elements. *Occupational mobility* is an individual's ability to change occupation by virtue of his level of acquired skill, whereas *industrial mobility* is an individual's potential, by virtue of his occupation, to be employed in more than one branch of industry. It would appear that few people attain their potential mobility, and 'policies to encourage industrial and occupational mobility on the scale desired must attempt to influence basic attitudes to change both by management and worker' (Ministry of Labour, 1966, p. 381). *Social mobility* is a term which is difficult to define. It is itself a multidimensional concept with education, training, character, ability and personality as important components. It is strongly linked with *geographical mobility* in cases where both family and community ties limit movement to minor locational changes within a limited area (Young and Willmott, 1962; Willmott and Young, 1967).

Therefore labour mobility is the ability, willingness and action of an individual to change employer, occupation and residence—or even to move into or out of the labour force. The effects of levels of labour mobility within a region can be to reduce the size of the labour force and alter its composition.

(*b*)   *The labour market.* Employers and employees together constitute the two basic elements of a labour market. It has been suggested that each employment unit in an area forms a distinct market for labour not only because it has an internal market amongst its existing employees but also because its location and package of rewards are unique. Hunter (1969) considers the local labour market to be an area 'in which there is a concentration of employment within which most resident workers can change jobs without changing their place of residence' (p. 42), and it is presumed that journey-to-work flows delimit the approximate spatial dimensions of the labour market.

(*c*)   *The journey to work.* In the short term, at the sub-regional level, the location of industry and markets can be assumed to be static, but the population is decentralizing to suburban locations. The fact that, in 1961, 36 per cent of the economically active population in England and Wales worked outside the local authority area in which they lived (Lawton, 1968) means that questions relating to the volume of journey-to-work movements and the extension of labour-supply areas are of immediate concern.

The implications of the spatial separation between residential and industrial land uses have not been clearly resolved. This has led to a variety of opinions which range from stating that manufacturing establishments can be located anywhere within an urban region and be assured of all the necessary labour force (Voorhees, 1965) to a consideration of direct accessibility between residential and industrial areas as one of the principal prerequisites in the determination of industrial location within urban areas (Chapin, 1965).

2   Local labour markets in the English East Midlands

(a) *Delimitation of labour market areas.* The underlying structure of journey-to-work flows can be determined using the graph theory approach of Nystuen and Dacey (1961), which is a procedure for grouping areas into a hierarchy of nodal regions on the basis of the magnitude and direction of inter-urban flows. The theory is based on three main concepts. A city is declared to be independent if its largest outflow is to a smaller city, where city size is defined as the sum of all inflows to each city in the network. The hierarchy has a structure of independent-subordinate relationships where, if city A is subordinate to city B and city B is subordinate to city C, then A is also subordinate to C. In order to extract the latent structure of inter-city functional associations, only the major outflow from each city is considered (concept 1), but total inflow to each city is used to arrange the cities in rank size (concept 2) to determine the direction of their association (concept 3).

The procedure was used by Nystuen and Dacey to delimit nodal regions where inter-city telephone calls were used as a surrogate for all inter-city contact. The process can be applied to journey-to-work flows, and its use will be illustrated in the delimitation of component labour markets within the Nottingham–Derby area in the English East Midlands. It will serve as a classification procedure to allocate each local government unit to a specific labour market, but it will not indicate how these markets were formed or how they function. Rural Districts are incorporated into the first stage of the analysis in order to present an overview of the labour market area as a whole before concentrating on the inter-urban links within the area.

The inter-area flows were subdivided into their male and female components prior to the construction of the origin–destination matrix based on journey-to-work flows for 1966. The analysis of the matrix reveals three distinct labour market areas (Fig. 5.1) centred upon Nottingham, Derby and Mansfield. Labour markets do not exist as discrete self-contained areas, and their delimitation will incorporate marginal areas on the periphery of the market which have well-developed connections with more than one market area. Concentrating on the Nottingham and Derby labour market areas it can be seen (Table 5.1) that there is a balance between the resident working population and the daytime working population, with only marginal net gains and losses through the interchange with other areas.

If the male labour markets are examined in a little more detail it can be seen that they are characterized by a three-tier structure,

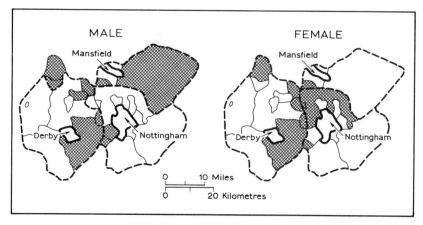

Fig. 5.1 1966 labour market areas (dashed boundaries) of Nottingham, Derby and Mansfield. The shaded areas have over ten per cent of their outflow to another market area.

Table 5.1 Nottingham and Derby labour market areas

	Nottingham	Derby
Male resident occupied population	201,410	127,510
Female resident occupied population	110,390	70,270
Male daytime occupied population	198,930	133,800
Female daytime occupied population	113,140	70,750
Net difference male	—2,480	6,290
Net difference female	2,750	480

Source: Registrar General (1968)

with the majority of local government areas located on the second rank of the hierarchy oriented towards the major centre (Fig. 5.2).

(*b*) *Evolution of labour market areas.* The structural evolution of the market areas can be illustrated with reference to the dominant inter-urban flows. The urban system can be depicted as a series of nodes with the workflows representing links between them. It is suggested (Cowan and Fine, 1969) that the development of nodes and links within such a system will follow a logistic curve. The system develops links not only as the number of nodes increases but also by the multiplication of links between the nodes already in the system.

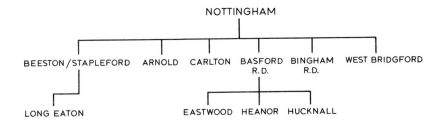

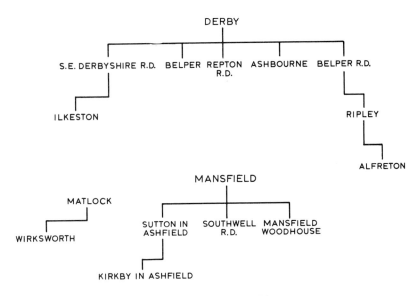

Fig. 5.2 Nodal labour market areas; males 1966.

Therefore at the point in time when the addition of nodes represents a marginal increase or halts altogether, the links will still continue to increase both in magnitude and number.

In the Nottingham–Derby urban system, the number of links increased from 62 in 1921 to 136 in 1951, but between 1951 and 1966 some contraction took place and the total number of links decreased to 115. The reduction may be attributed to the slow recovery of housing development in the immediate postwar period, which imposed a restriction on residential mobility and was manifest in a temporary increase in inter-area journey-to-work movement. If a severe constraint is imposed on the definition of a link so that the outflow from one node to another must constitute at least ten per cent of the total outflow from the origin node, then the basic form of the system will be evident. Between 1921 and 1951 (Fig. 5.3) the number of such links increased from 20 to 38 and then declined to 36 in 1966. It would appear that the major development took place between 1921 and 1951, and that subsequent development will have been concentrated in the volume of movement along the links.

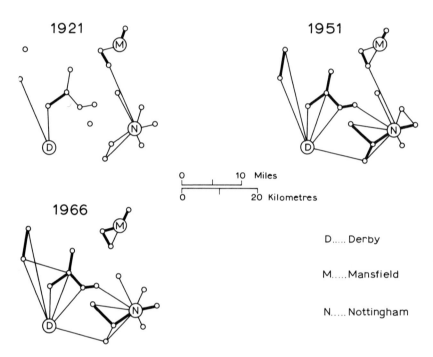

Fig. 5.3 Inter-urban linkage development in the Nottingham, Derby and Mansfield area. Thick lines represent a two-way link.

(*c*) *Urban self-containment.* Having defined the labour market areas and briefly examined their evolution, it is relevant to ascertain the degree of self-sufficiency of the urban centres in terms of employment for their resident populations. This can be achieved by using an 'Index of Independence' (Thomas, 1969), which is expressed as the ratio of daily work-trips within a local authority area to the total volume of cross-boundary movement into and out of that area. The higher the Index value, the greater the degree of self-containment.

The general pattern has been towards a decrease in the Index value between 1951 and 1966 (Table 5.2). This indicates that the urban areas in the Nottingham–Derby district are moving towards a

Table 5.2   Index to independence: urban areas

Area	1951	1966
Alfreton U.D.	1·00	0·75
Arnold U.D.	0·38	0·34
Beeston and Stapleford U.D.	0·65	0·63
Belper U.D.	0·93	0·98
Carlton U.D.	0·50	0·42
Derby C.B.	1·60	1·03
Heanor U.D.	0·81	0·81
Hucknall U.D.	1·18	0·65
Ilkeston M.B.	0·91	0·55
Kirkby in Ashfield U.D.	0·66	0·56
Long Eaton U.D.	0·90	0·85
Mansfield M.B.	0·78	0·70
Mansfield Woodhouse U.D.	0·32	0·32
Nottingham C.B.	1·96	1·51
Ripley U.D.	0·63	0·56
Sutton in Ashfield U.D.	1·44	1·01
West Bridgford U.D.	0·23	0·30

Source: Registrar General (1954, 1956, 1968)

greater degree of interdependence. The changes in index value are related to the allocation of population and employment within the area. The working population resident in the urban centres in 1966 was 433,100, of which 140,180 (32 per cent) *worked* outside their area of residence: 61,790 (58 per cent) of the males and 28,760 (81 per cent) of the female working population commuted between the urban centres. The movements were oriented on Nottingham and Derby which accounted for 45 per cent and 56 per cent of the respective male and female inter-urban movement. Between 1951 and 1966 both centres declined in population, despite an extension of Nottingham's administrative boundary, but had a 22·4 per cent share of the employment increase located in their labour market areas.

At the inter-regional scale, disparities in the location of population and employment opportunities gives rise to unemployment and inter-regional migration, whilst at the intra-regional scale it produces increasingly complex journey-to-work movements. Close proximity of urban settlements within a sub-region presents the opportunity for localized specialization, with resulting scale economies. This can provide the necessary set of conditions for the development of a dispersed city (Stafford, 1962), and the urgency for an integrated planning policy amongst local authorities in such an area becomes all the more acute.

IV Labour supply and labour turnover

1   *The turnover process*

One of the prerequisites for continuing economic growth within a region is a substantial and diverse supply of labour. Evidence has been produced to show that towns are becoming increasingly inter-dependent and this would suggest a greater mobility of the population and a consequent extension or strengthening of labour market areas. This has great significance for industrial location at the micro-scale. If a firm's labour supply area is extending over time, then an increased proportion of the population resident within an area become potential employees, but the increase in mobility levels will also intensify inter-firm competition for labour.

The classic economic theory of competition assumes that changes in wage level will produce an adequate supply of labour just sufficient to meet the increased demand; that labour will move from low to higher paid jobs; and that the industrial wage structure will allocate labour amongst the competing industries. But these assumptions cover an intricate network of causal factors the inter-relationship between which is not fully understood. The situation is complicated by the occupational structure within an industry, the meaning of wage, the availability of overtime, bonus schemes, fringe benefits and a firm's hiring policy. There is pressure on management to turn to non-wage alternatives in order to secure labour for fear of creating a wage spiral in both the internal and external labour markets (O.E.C.D., 1968). Moreover, it is possible to have wage differentials within an industry and within a single occupation caused by devia-tions from nationally agreed wage levels determined by plant nego-tiations. Although positive relationships have been established between migration and changes in earnings (O.E.C.D., 1965; Greenwood, 1969), there is no substantive body of evidence to sug-gest that the same processes operate at the local level, where the information flow between employers and employees plays a major part in precipitating inter-firm movement.

2   *Differences in turnover rates among industries*

Underlying the process of labour turnover are changes in demands made by the different sectors of industry. It is essential to determine the factors behind different turnover rates to place the geographical methodology within a wider perspective. For the manufacturing sector as a whole in the U.K. the period 1950–65 saw engagements exceed discharges during the periods 1953-5 and 1959-65; this applies in both the male and female sectors. A distinction lies, however, in the higher turnover rate for the female labour force and its more pronounced excess of engagements over discharges.

When the fourteen major manufacturing orders are grouped together on the basis of an excess or deficit of engagements over discharges for the period 1950-65, they fall into two groups; one with an expanding labour force (vehicles; metals n.e.s.; engineering and electrical goods; chemical and allied industries; paper, printing and publishing; bricks, pottery, glass and cement; food, drink and tobacco) and one with a contracting or static labour force (timber and furniture; leather goods and fur; clothing and footwear; textiles; metal manufacturing; marine engineering; other manufacturing). The application of national trends to the regional and sub-regional scale must be viewed with caution, and local changes may only be commensurate with national changes for the declining industries because the expanding industries exhibit a marked degree of regional differentiation. In addition, the activities of large employers of labour within an area may govern the local situation for labour demand and supply and this can lead to local employment levels anomalous to the national level (Sant, 1967), although there has been a detectable convergence of regional industrial structures on the national structure as measured by a coefficient of specialization (Cunningham, 1969).

Industries, and hence firms with high levels of both discharge and engagements which indicate a heavy rate of labour wastage, may incur substantial additional expenditure in training costs and induction programmes. High labour-wastage rates are symptomatic of some industries, and reflect either the nature of the occupational structure or numerous other factors, both internal and external to the industry or firm, which induce movement (Samuel, 1969; Silcock, 1954). For example, there is a negative relationship between national discharge and engagement rates and the national level of unemployment. Periods of relatively high unemployment are accompanied by a lowering of the rate of both discharges and engagements. At such times movement between jobs and movement into and out of the labour market will be at a minimum. This is the individual's response to a worsening employment situation where movement out of secure employment may mean moving into a state of unemployment. The analyst must be mindful of these overriding considerations when examining the spatial processes external to the firm which increase the total information level on labour movement.

3   *The household's decision-making process*

Factors other than those just mentioned have thus to be taken into account when explaining labour movement. One of the characteristics of movement within a local labour market is the opportunity afforded to change workplace location without an attendant change in residential location. Similarly, residential moves need not be associated with changes in job location and therefore may lead to the extension of a firm's labour supply area (Goldstein and Mayer, 1964).

Kain (1962) has suggested that households substitute journey-to-work expenditure for residential site expenditure, the substitution depending primarily upon household preference for low density residential sites. Space preference is determined by family size and constrained by household income and price per residential unit. He then introduces modal choice by stating that once space preference has been decided upon, the second decision is whether or not to purchase an automobile. This depends upon the residential site decision, is conditioned by the available public transport service and is again constrained by income. The length of the journey to work is then determined by the residential location decision and the mode choice of and, with reference to trip costs, is constrained by income.

The implications of this model are concerned with the relationship between age, income, sex and house tenure to travel mode and travel time expended on the journey to work. In addition there is the important question of changes in modal split composition, especially the transfer from public to private transport.

It has been suggested that commuting and residential migration decisions are interdependent, so that an increase in costs brought about by a change in either workplace or residential location within an area may be considered to be more desirable than the costs incurred by migration into another local labour market area (Lakshman, Polese and Wolpert, 1969).

4   *Residential movement and labour supply*

Residential movement follows a constant and well-defined cycle, which begins with newly married couples leaving their nuclear family units and reaches a peak when families with young children move because of increased space demands. From this point movement steadily declines until old age produces a relatively immobile population (Cullingworth, 1965). Because the intensity of residential movement is concentrated within specific age groups, there may be differences in journey-to-work trip lengths between age groups. Thompson (1956) found that in West Virginia middle-aged commuters had the highest average trip lengths. However, this conclusion was not supported in a survey of firms in the U.K. (Rankin, 1971) where the highest average trip lengths were found in the 15–30 years age category.

A household must attain a certain income level before residential movement can be contemplated, unless it is movement enforced through the implementation of urban renewal policies. Cullingworth (1965) found that residential movement was related to specific income groups. There are many other reasons underlying residential change. In accordance with the movement cycle, movement is initially precipitated by marriage, which either accompanies or precedes house purchase. Evidence that the second sequence is probably the more typical is provided by C. J. Thomas (1966) who indicated that tenants on public housing estates saved during the early years of their occupancy and then moved to private owner-occupier housing.

In a Nottingham survey (Rankin, 1971) the major reasons given for employees moving were marriage, house purchase and transfer from a private rented establishment to a public housing estate. Movement which incorporated a conscious desire to move either nearer to or further away from the workplace was not highly represented. The suggestions that households wish to live further away from the workplace and are willing to undertake lengthy worktrips (Humphreys, 1965) may not be universally verified. This type of movement may either reflect the life style of some sections of the labour force or result from the limited availability of job opportunities in specific localities.

In the Nottingham survey employees were subdivided on the basis of marital status and analysed for differences in trip lengths. No significant difference in trip length was established between married and single employees, whether male or female. Therefore, whilst marriage promotes residential movement within a local labour market, it would appear not to affect trip length. Conversely Thompson (1956) showed in West Virginia that married employees commuted the furthest distance. The conditions and attitudes affecting residential movement and the consequences of that movement for labour supply must therefore be evaluated within the context of the area under investigation.

If we turn to the operation of the housing market with specific reference to the location of public housing and private development, the importance of residential movement to intra-urban labour availability may become clearer. The trend in England and Wales during the postwar period has been for the number of private dwellings to increase at a steady rate, whilst the completion rate for public dwellings declined during the 1950s and increased during the 1960s. Over the whole period 1951–69 the total number of buildings completed was 5,179,581, of which the public and private shares were respectively 50·3 per cent and 49·7 per cent.

The location of new private and public housing will affect the house-tenure status of the population resident in each local authority area. For England and Wales the change between 1961–6 was towards an increase in both owner-occupier (42·3 per cent to 46·7 per cent) and public housing tenant (23·7 per cent to 25·7 per cent) status with an accompanying decrease (27·8 per cent to 22·5 per cent) in the private rented sector of the housing market. The movement towards private ownership was most marked in the smaller towns and rural areas. This pattern of development is to be expected as the availability of land for building in the cities declines, as urban renewal projects rebuild decaying areas at lower densities, as per capita incomes rise, and as car ownership increases. Urban populations are decentralizing from the cities to the suburbs.

The location and relocation of population is dependent upon the location of both public and private housing projects. Although this statement may appear to be axiomatic, it is important because it affects individual residential location decisions. Prospective public tenants may be even more restricted because of the residential qualifications essential for a place on a housing allocation list and the (often lengthy) period of time before a house is allocated. Cullingworth (1965) found a high proportion of council tenants whose only reasons for residing in a particular location were that the house had been allocated or was the only one available. Similarly the Toothill Report (1961) indicated that although council tenants seemed able to move quite easily within local authority areas, the methods used in allocating houses discouraged people from moving to obtain better employment.

The relevance of house tenure to the interpretation of some journey-to-work movements is clearly stated by Pahl (1967) when writing about commuters from the London Metropolitan rural-urban fringe.

> For members of the working class who commute to work out of the area, space is a constraint which has to be overcome and the cost of doing so is an added economic burden to the family. Not all the working class choose to live in a rural area but they are forced to do so by a society which allocated council houses by area of residence and not place of

present employment. By contrast, for a section of the middle class, space is valued as an amenity which should be preserved. The economic burden of crossing it in order to reach urban employment and amenities is an accepted concomitant of their way of life. Such people choose to live in an area where working-class people may be forced to live and the residential pattern is a resolution of the two forces (p. 239).

This viewpoint is supported by Pennance and West (1969), who indicate that peripheral residential development will only attract a few people from core areas because of the transport cost. Therefore low-income households are location orientated and high-income households are price orientated.

V The female labour force

The postwar period has witnessed an unprecedented increase in female participation in the labour force. In Great Britain in 1947 there were 6·6 million employed females but by 1969 this had increased to 8·6 million and constituted 37 per cent of the total labour force. The increase in female participation is mainly due to the rise in activity rates of the married sector. In 1950 41 per cent of the female labour force was married and by 1969 this had risen to 59 per cent. Despite this increase, married females are one of the manpower resources which are still under-utilized and who could make a valuable contribution to the future national labour requirements (Klein, 1965). By 1981 it is expected that over half of the U.K. labour force will be female (Lawton, 1969).

1 *Factors influencing female participation*

The initial conditions are set by the prevailing social attitudes to married female employees; and the changes in attitude are reflected both in the hiring policies of individual employers and within the family unit itself, where to an increasing extent the female fulfils the dual role of domestic and employee. Within this attitudinal framework, the specific processes which constrain female participation in the workforce must be identified and then examined to determine their spatial component. The factors examined depend upon the scale of the investigation.

Gordon (1970) examined regional differences in activity rates in 1966 for standard statistical areas in Great Britain. Using a multiple regression technique with age-adjusted female activity rates as the dependent variable, Gordon found that variations in the activity

rates were accounted for by inter-regional differences in unemployment levels, rurality of the population and industrial structure.
Although the increase in activity rates between 1961–6 varied from
region to region, there was a measurable tendency towards a conversion of rates between the regions. Corry and Roberts (1970), in an
analysis of female activity rates and unemployment levels, also
established a negative relationship between the two variables and
indicated that the fluctuations were likely to be concentrated in the
25–44 years age group.   The regional distribution of female activity
rates in 1966 reflected the distribution of population and unemployment opportunities within the country: activity rates in excess of 40
per cent were mainly achieved in the South East–North West belt
and decreased outwards from this zone.

Taylor (1968), investigating sub-regional variations in female
activity rates, stated that the demographic characteristics of the
population, the industry mix, male earnings levels and local traditions were the four major factors regulating activity rates. Taylor
concludes that if the manufacturing sector is to achieve a marked
growth in output it will require additional labour and this will be
principally drawn from the currently inactive married females.
The distribution of female labour reserves has not only pronounced
regional but also sub-regional variations in intensity. These spatial
variations in labour availability have implications for government
regional industrial location policy and also for the location of
industry at the micro-scale.

Carmichael (1971), in a study of married female activity rates
within the Greater Auckland area, found that the inter-suburban
differences in activity rates were negatively associated with two
variables which together accounted for 81 per cent of their variation.
The variables were an index of transience which expressed the number of rented, loaned and temporary dwellings as a percentage of all
private dwelling and was designed as a proxy variable to measure
economic stress on newly married couples seeking a home, and a
family status variable which measured households with children in
the 0–4 years age category. The distribution of residual values
showed marked areal variations, with the positive residuals and near
zero residuals situated in suburbs on the isthmus in the core of the
urban area and the negative residuals associated with the coastal
suburbs most distant from the majority of employment locations.
The distribution would appear to reflect not only demographic
differences between the residential areas but also a sensitivity to
distance and travel time between the employment and residential
areas.

Because of the low level of female car ownership a reliable and efficient public or works service is a prerequisite to attract female labour, especially married female labour, from locations some distance away from the factory site. Convenience of hours will also be important, so innovations such as twilight shifts and variable starting times for shifts would be extra attractions for the married female. For unskilled or semi-skilled work the female not already in the labour market may have a range of potential job opportunities to choose from, and recommendations from friends or relatives can play an important part in the final decision. In addition the atmosphere within the factory and job satisfaction can be powerful factors in the retention of labour, often more important that the financial remuneration itself.

2   *Female labour supply areas—an example of two firms*

Two examples will be examined to illustrate how trip time, trip length, travel mode and residential location affect the configuration of labour supply areas for individual firms depending upon the location of the firm itself. Two firms have been chosen from the Nottingham, U.K., area, one firm centrally located in an area adjacent to the C.B.D. employing 1,400 females, the other firm chosen for its suburban location employing 1,900 females. The female labour force at both firms are either engaged in administration or assembly work.

An examination of trip length indicates that the centrally located firm has the more extensive catchment area (Fig. 5.4A) with an

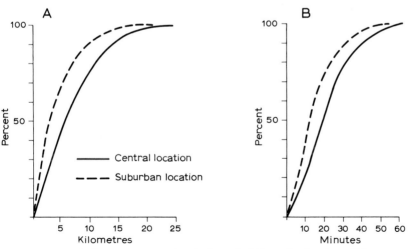

Fig. 5.4 Female journey-to-work trip length (A) and trip time (B).

average trip length of 6·2 km in contrast to 3·8 km for the suburban firm. The disparity in supply-area size is however only accompanied by a minor difference in trip time (Fig. 5.4B); average trip time for the central location is 21 minutes and for the suburban location 15 minutes. The similarity in trip time can be explained with reference to the modal split of the two sets of employees. 54·6 per cent of the females at the centrally located firm travelled by bus and only 17 per cent walked to work whilst for the suburban firm 36 per cent travelled by bus and 33·7 per cent walked. The benefit derived by the central location comes from its adjacency to major intra-city transport routes and public transport services. Although modal split composition is subject to considerable areal variations and average figures for extensive areal aggregates frequently bear no relationship to individual studies (Hammond, 1968), the accessibility level means that the absence of a suitable quality or supply of labour near to a site can be offset by good access. The compact nature of female labour supply areas suggests that there has been little extension during the post-war period, and that because of the constraints of domestic duties and time, female labour is still relatively immobile.

VI Conclusion

An adequate supply of labour is fundamental to the implementation of a country's economic growth policies. Unfortunately, differences in the spatial distribution of manpower and employment opportunities create hardship and unemployment in some regions and congestion and labour shortages in others. At the sub-regional scale, even in areas where there is a plentiful supply of labour, firms by virtue of their location can experience difficulties in obtaining the necessary work-force. Changes in the spatial structure of economic organization can be achieved through the operation of market forces constrained by the framework of government and local authority planning policies (see Ch. 7). The present chapter has focused attention on some of the problems and indicated where the application of geographical concepts can contribute to manpower planning policies.

References

ALLARDT, E. (1969): Aggregate analysis: the problem of its informative value, *in* M. DOGAN and S. ROKKAN, editors: *Quantitative ecological analysis in the social sciences*. Cambridge, Mass.

BARLOW REPORT (1940): *Report of the Royal Commission on the distribution of industrial population*. London.

BERRY, B. J. L. and E. NEILS (1969): Location, size and shape of cities as influenced by environmental factors: the urban environment writ large, *in* H. S. PERLOFF, editor. *The quality of the urban environment*. Baltimore.

BRENIKOV, P. (1967): Geography and planning, *in* R. W. STEEL and R. LAWTON, editors. *Liverpool essays in geography*. London.

BRITTON, J. N. H. (1967): *Regional analysis and economic geography*. London.

BUNGE, W. (1962): Theoretical geography. *Lund Studies in Geography, Series C*, 1.

BUSWELL, R. J. and E. W. LEWIS (1970): The geographical distribution of industrial research activity in the United Kingdom. *Regional Studies* 4, 297–306.

CARMICHAEL, G. A. (1971): *Women at work: a geographical analysis of Auckland's female labour force*. Unpublished M.A. thesis, University of Auckland.

CHAPIN, F. S. (1965): *Urban land use planning* (second edition). New York.

CHISHOLM, M. (1966): *Geography and economics*. London.

CLARK, C. (1966): Industrial location and economic potential. *Lloyds Bank Review* 82, 1–17.

CORRY, B. A. and J. A. ROBERTS (1970): Activity rates and unemployment: the experience of the United Kingdom 1951–66. *Applied Economics* 2, 179–201.

COWAN, P. and D. FINE (1969): On the number of links in a system. *Regional Studies* 3, 235–42.

CULLINGWORTH, J. B. (1965): English housing trends. *Occasional papers in Social Administration* 13. London.

CUNNINGHAM, N. J. (1969): A note on the 'proper distribution of industry'. *Oxford Economic Papers* 21, 122–7.

DYCKMAN, J. W. (1963). Transportation in cities. *Scientific American* 213, 162–74.

ESTALL, R. C. and R. O. BUCHANAN (1966): *Industrial activity and economic geography* (second edition). London.

GODDARD, J. (1968): Multivariate analysis of office location patterns in the city centre: a London example. *Regional Studies* 2, 69–85.

GOLDSTEIN, S. and K. MAYER (1964): Migration and the journey to work. *Social Forces* 4, 472–8.

GORDON, I. R. (1970): Activity rates: regional and sub-regional differentials. *Regional Studies* 4, 411–24.

GREEN, F. H. W. (1953): Community of interest areas in Western Europe: some geographical aspects of local passenger traffic. *Economic Geography* 29, 283–98.

GREENWOOD, M. J. (1969): The determinants of labour migration in Egypt. *Journal of Regional Science* 9, 283–339.

GRIME, E. K. and D. N. M. STARKIE (1968): New jobs for old: an impact study of a new factory in Furness. *Regional Studies* 2, 57–67.

HAMMOND, E. (1968): *London to Durham: a study of the transfer of the post office savings certificate division*. Durham.

HARRISON, C. R. (1967): *Recent changes in the industrial geography of Greater Leicester, with particular reference to new industries*. Unpublished M.A. thesis, University of Nottingham.

HEMMENS, G. C. (1966): *The structure of urban activity linkages.* Chapel Hill.

HIRSCHMAN, A. O. (1958): *The strategy of economic development.* New Haven.

HOOVER, E. M. (1963): *The location of economic activity.* New York.

HOUSE, J. W. and B. FULLERTON (1960): *Teesside at mid-century: an industrial and economic survey.* London.

HOUSE, J. W. and E. M. KNIGHT (1966): People on the move: the south Tyne in the sixties. *Papers on migration and mobility in North Eastern England* **3.** Newcastle.

HUMPHREYS, G. (1965): The journey to work in industrial South Wales. *Transactions of the Institute of British Geographers* **36,** 85–96.

HUNTER, L. C. (1969): Planning and the labour market, *in* S. C. ORR and J. B. CULLINGWORTH, editors: *Regional and urban studies.* London.

KAIN, J. F. (1962): A multiple equation model of household locational and trip making behaviour. *Rand Corporation Report* **RM-3086 FF.**

KEEBLE, D. E. (1968): Industrial decentralization and the metropolis; the north-west London case. *Transactions of the Institute of British Geographers* **44,** 10–54.

(1970): The movement of manufacturing industry—comments. *Regional Studies* **4,** 395–7.

KLEIN, V. (1965): *Britain's married woman workers.* London.

KUKLINSKI, A. R. (1970): Regional development, regional policies and regional planning: problems and issues. *Regional Studies* **4,** 269–78.

LAKSHMAN, Y., M. POLESE and J. WOLPERT (1969): Interdependence of commuting and migration. *Proceedings of the Association American Geographers* **1,** 163–8.

LAWTON, R. (1968): The journey to work in Britain: some trends and problems. *Regional Studies* **2,** 27–40.

(1969): Putting people in their place. *Geographical Magazine* **41,** 927–36.

LUTTRELL, W. F. (1962): *Factory location and industrial movement,* vol. I. London.

MCCRONE, G. (1969): *Regional policy in Britain.* London.

MINISTRY OF LABOUR (1966): *Ministry of Labour Gazette,* 381. London.

MYRDAL, G. (1957): *Economic theory and underdeveloped regions.* London.

NEEDLEMAN, L. (1965): What are we to do about the regional problem? *Lloyds Bank Review* **75,** 45–58.

NYSTUEN, J. D. and M. F. DACEY (1961): A graph theory interpretation of nodal regions. *Papers, Regional Science Association* **7,** 29–42.

ORGANIZATION FOR ECONOMIC COOPERATION AND DEVELOPMENT (1965): *Wages and labour mobility.* Paris.

(1968): *Employment stabilization in a growth economy.* Paris.

PAHL, R. E. (1967): Sociological models in geography, *in* R. J. CHORLEY and P. HAGGETT, editors. *Models in Geography.* London.

PENNANCE, F. G. and W. A. WEST (1969): Housing market analysis and policy. *Hobart Papers* **48.**

PRED, A. (1966): *The spatial dynamics of U.S. urban-industrial growth, 1800–1914: interpretive and theoretical essays.* Cambridge, Mass.

RANKIN, D. G. (1971): *Journey to work movements and labour supply areas in the Nottingham–Derby district with specific reference to industrial firms.* Unpublished Ph.D. thesis, University of Nottingham.

REGISTRAR GENERAL (1954): *Census 1951, County Reports for Nottinghamshire and Derbyshire.* London.

(1956): *Census 1951, Report on usual residence and workplace.* London.

(1968): *Sample census 1966, workplace and transport tables part 1.* London.

REYNOLDS, L. G. (1951): *The structure of labour markets.* New York.

RICHARDSON, H. W. and E. G. WEST (1964): Must we always take work to the workers? *Lloyds Bank Review* **71**, 35–48.

SALT, J. (1969): Post-war unemployment in Britain: some basic considerations. *Transactions of the Institute of Britain Geographers* **46**, 93–103.

SAMUEL, P. J. (1969): Labour turnover? towards a solution. *Institute of Personnel Management Publication.* London.

SANT, M. E. C. (1967): Unemployment and industrial structure in Great Britain. *Regional Studies* **1**, 83–91.

SILCOCK, H. (1954): The phenomenon of labour turnover. *Journal of the Royal Statistical Society, Series A.* **117**, 429–40.

SMITH, D. M. (1969): *The North West.* Newton Abbot.

(1970): On throwing out Weber with the bathwater: a note on industrial location and linkage. *Area* **3**, 15–18.

STAFFORD, H. A. (1962): The dispersed city. *Professional Geographer* **14**, 4–6.

STEWART, J. Q. and W. WARNTZ (1958): Macrogeography and social science. *Geographical Review* **48**, 167–84.

TAAFFE, E. J., B. J. GARNER and M. H. YEATS (1963): *The peripheral journey to work: a geographic consideration.* Evanston.

TAYLOR, J. (1968): Hidden female labour reserves. *Regional Studies* **2**, 221–31.

TAYLOR, M. J. (1970): Location decisions of small firms. *Area* **2**, 51–4.

THOMAS, C. J. (1966): Some geographical aspects of Council housing in Nottingham. *East Midland Geographer* **4**, 88–98.

THOMAS, R. (1966): Industry in rural Wales. *Welsh Economic Studies* **3**.

THOMAS, R. (1969): London's New Towns: a study of self-contained and balanced communities. *P.E.P. Broadsheet* **510**.

THOMPSON, J. H. (1956): Commuting patterns of manufacturing employees. *Industrial Labour Relations Review* **20**, 70–80.

TOOTHILL, J. N. (1961): *Inquiry into the Scottish Economy 1960–61.* Edinburgh.

TOWNROE, P. M. (1970): Industrial linkage, agglomeration and external economies. *Journal of the Town Planning Institute* **56**, 18–20.

TULPULE, A. H. (1969): Dispersion of industrial employment in the Greater London area. *Regional Studies* **3**, 25–40.

VOORHEES, A. M. (1965): Factors influencing growth in American cities. *Highway Research Record* **24**, 83–95.

WALKER, L. G. (1967): The delimitation of urban areas. *Leeds School of Town Planning Research Publication* **1**.

WILLMOTT, P. and M. YOUNG (1967): *Family and class in a London suburb.* London.

WOOD, P. A. (1969): Industrial location and linkage. *Area* **2**, 32–9.

YOUNG, M. and P. WILLMOTT (1962): *Family and kinship in East London* (revised edition). London.

6 Marketing

John A. Dawson

The businessman making decisions concerning the expansion of his selling operations examines many problems. Where in a region should he site his new warehouses? How can forecasts be made of the acceptance of his new product through a region? How will an advertising campaign vary in relation to the differences in social structure in a region? What type of retail organization will buy particular products in town A and how will this differ in town B? Many such questions have to be asked and answered, and the problems they pose are of interest to the geographer in that they arise from the effects of *place* on the marketing operations; moreover, the solutions adopted necessarily affect man's everyday environment.

The trading instinct in man and the trade function in society are apparent in the earliest of civilizations (Polanyi, Arensberg and Pearson, 1957). Traders from Mesopotamia, in the second and third millennia B.C., exchanged articles with people in Northern Europe. In general, however, consumption in these early societies was met by local production, with producer and consumer often one and the same person. As the economic structures in Europe became more complex so the gap between production and consumption widened. This gap was apparent not only in terms of changes in social organization, such as the division of labour, but also in the spatial separation of the producer and consumer. Sections of society specialized in the production of particular goods, for example with the establishment of handicraft manufacturing and the development of guilds in the medieval city. In parallel with this increasingly specialized production, the demand for goods became more varied. Groups of consumers in society had different mixes of demands for goods; demand thus developed *cross-sectional* variety. As parts of society became agglomerated and other parts remained dispersed, so demand began to vary between town and country and demand developed *spatial* variety. With increased specialization of production on the one hand and greater variety in demand on the other, so the equation of demand and production became more important and more complex. The activities in this equation close the space, time, quantity, and ownership gap between, in general, production in large quantities at unique points and demand for small quantities at many points. Furthermore it is these equalizing activities which now form the complex marketing system of high-level consumption economies. In some parts of the world the marketing system remains far simpler, with the producer and the consumer less distinct and more disparate in function. Within any present-day space economy, marketing is a function of the specialism of production meeting the variety in demand.

The meaning of marketing is considered further in the first section of this chapter and some basic problems to which the geographer may address himself are defined. The composition and interrelationships of the marketing system is considered in more detail in the second section. The search for theory in marketing has concerned both geographers and economists. The introduction of geographical considerations into some marketing theories is discussed in Section III. From theory we move to case studies in Section IV, with an outline of some of the spatial elements of the marketing system. Finally, mention is made of the variety of studies which investigate the part played by the planning of marketing systems in the overall process of economic and social development. Many of these planning studies are the work of geographers, whilst others are the work of research teams which include geographers.

I Definitions and problems

1   *The scope of marketing*

Marketing is concerned with the transfer of goods from the producer to the consumer. Thus it is concerned with the processes of physical movement, storage, and ownership of goods. As these processes operate, so goods change their location. This basic introduction of the space component immediately introduces the geographer to the scene. The separation of consumption and production, which is the reason for marketing operations, is essentially a spatial idea and hence the problems of resolving this separation have a major spatial component. For solutions to these problems the marketing scientist is turning increasingly to the geographer and to methods of geographical analysis.

The enormous breadth of these problems is immediately apparent, for the concept of marketing relates to all types of commodity, from steel plate to shoes. The problems involved in directing flows from producer to consumer vary with the product; the problems differ and the solutions differ. Add to this the differences in each of these composite markets—for example, the regional variations in demand for different types of shoes—and the problems in the marketing system of a high-level economy become immense.

2   *Problem-solving in marketing*

(*a*) *The input-output approach.* In an attempt to introduce order into the definition of these problems, input-output analysis has been used. This approach allows for the study of the spatial component and its particular problems in the marketing system.

Input-output analyses, in their simplest form, are usually concerned with three basic tables. The first is a transaction table showing how the demand from consumers of finished goods influences the flows of raw materials, semi-finished goods, and requisition services. The second is a table which shows by industry the inputs required to produce a fixed worth (e.g. one dollar) of output. The third is a table showing by industry the total inputs (direct and indirect) required to produce and deliver to the final demand a fixed worth of output. The production of such tables and their analysis is a complicated and difficult operation. Geographers and spatial economists in recent years have attempted to produce such tables for regions (Hewings, 1971; Miernyk, 1970; Isard, Langford and Romanoff, 1967), and it seems likely that similar tables will eventually appear for towns.

The implications of input-output techniques to the practising market scientist are varied. This method involves breaking down the market, then rebuilding it showing all the interrelationships. As a result, patterns of sales appear which were previously submerged, the effects of change in demand may be foreseen and a comparison of an individual firm against the industry in general may show up markets in which the firm has taken little interest hitherto. Additionally, a detailed study of transport costs can be undertaken. In all these instances, the problem is to market a product efficiently. With the growing spatial and technological divergence between production and consumption, the marketing component in final prices is getting higher. Only by efficient marketing, and by seeking (even if not finally adopting) the optimum marketing pattern, can this marketing cost element be reduced.

(b) *The geographer-planner's approach.* Many of the specific problems in marketing are almost wholly spatial, and thus the geographer is in a position to apply his methods to the problems of the marketing scientist. The planners involved with French regional policy are currently concerned with the problem of increasing efficiency in the wholesale sector. The Fourth and Fifth National Plans advocate a pattern of regional wholesale markets. Some have been developed and Davy (1967) has studied the one at Nîmes. The location of such markets and the analysis of their viability are instances when the geographer can play an active role in applying his subject.

It is not only the marketing scientist who is interested in seeking an efficient marketing framework. Since many of the market processes produce a visible effect on the landscape by creating specialized land uses and by generating flows of people and goods, the economic and physical planner is deeply concerned with obtaining efficient marketing patterns. In this the geographer frequently has a voice, not least because the geographer-planner's approach may well consider more social goals than that of the marketing man.

II Marketing systems

1 *Composition of the system*

The very size and complexity of the elements, patterns, and interrelationships in marketing means that it is vital to consider the marketing processes as systems and subsystems. The equation of production with consumption when both vary through geographical space means that marketing is essentially an environmental system.

The system, furthermore, has landscape effect whether, for example, through the pressure for advertising space along roads, the location of service centres in newly-settled lands, or petrol filling-station location. Marketing is a complex, composite, open system.

> Marketing systems are large, complex, and intricate. They are large in several dimensions: in dollar amount, volume of goods handled, number of people employed, and quantity of components. They are complex in the number and types of elements and the possible variations within each element . . . There is negligible automatic adjustment in marketing systems. Since they do not adjust automatically, marketing management must plan to audit and to adjust the marketing system to better fit the wants, needs, opportunities of a changing market place. Managing marketing systems, therefore, is a difficult and demanding task. (Lazer and Kelley, 1962, pp. 684–5.)

The basic product movements of the marketing system are summarized and generalized in Fig. 6.1. Articles are transported from the production point, are then assembled, sorted, reassembled, transported and displayed by one or several *middlemen* (wholesalers, brokers, agents, retailers, etc.), and finally the article is consumed. The consumers themselves are relevant to the product flow within the marketing system for, by their demands, they influence the middlemen's operations. Regional demand patterns are reflected in regional concentrations of shop types and regional variations in retail structure (see Section IV.2). The marketing system also stretches back into the production units, for within the production unit the whole marketing process begins. Market research and market planning, for example, are particular functions which are carried out by the production unit or by agencies called in by the production unit. Vital to the marketing system is the movement of goods. In the simple scheme of Fig. 6.1 transport of articles occurs three times. Each time, however, the article has changed in form and occurs in association with different articles. Many of the processes in the marketing system are dependent on spatial transfers and the overcoming of the friction of space. Recent studies by geographers interested in the problems of transport have been concerned with the effectiveness of roads or railways in the distributive system (Kansky, 1963; Horton, 1968). Such geographical studies are of considerable value in the management of marketing systems.

Fig. 6.1 represents a considerable simplification of product movements within the marketing system, but none the less it serves to show the general nature of the channels of marketing between the producer and the consumer. One of the simplest channels is producer—private wholesaler—private retailer—consumer, involving

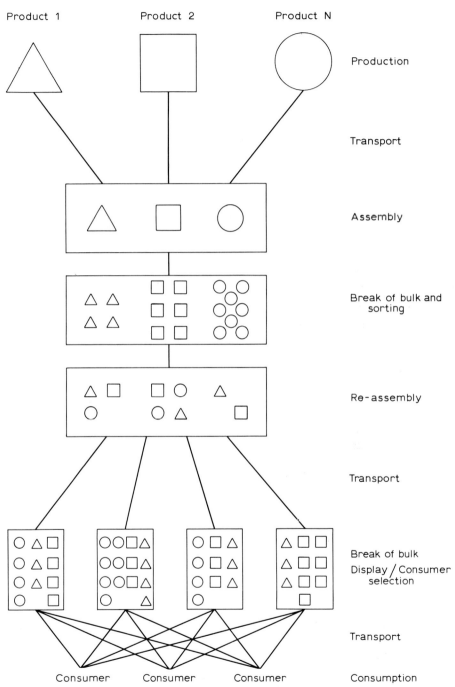

Fig. 6.1 The nucleus of the marketing system. After Fisk (1967).

two middlemen. More complex channels often result from vertical integration by producers and horizontal integration by middlemen. In many cases particular goods have individual channel structures.

2   *Market systems and economic wealth*

In general, most of the market processes operate within the so-called tertiary sector of the economy. A thesis of many geographers and economists is that growth in the tertiary sector, and hence increased complexity of marketing channels, is related to overall economic growth. In employment terms the marketing system in each of the countries of Western Europe accounts for between 40 and 50 per cent of the active labour force. Table 6.1 summarizes the importance of marketing to the economies of several European countries. The values listed probably underestimate the importance of marketing employment for many employees in these activities, for example, marketing managers in industrial concerns, are classified with particular industries. From the table the difference between Northern and Southern Europe is notable; and on a wider international scale even greater differences occur (Goldman, 1963; Mintz, 1956).

Table 6.1   Estimates of the percentage of civilian employment in commercial, transport and service activities

Selected European Countries	1960	1965	1970
United Kingdom	47	49	51
Denmark	40	41	48
France	39	41	45
West Germany	38	39	43
Norway	43	46	50
Finland	32	37	45
Netherlands	46	48	52
Switzerland	39	39	40
Spain	26	31	33
Portugal	28	30	32
Italy	29	31	35

Sources: Census data and O.E.C.D. labour force statistics

Within nations inter-regional variation is considerable, with some areas much more dependent on marketing for their economic wealth. The variation can also be seen amongst towns, and Fig. 6.2 indicates

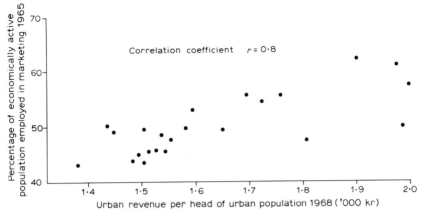

Fig. 6.2 The relationship between employment in marketing and potential for urban development in Danish towns with over 20,000 population—1965. *Sources:* Danish census of population 1965; Statistiske Eferretninger, 1970 (20), 1971 (2).

the extent (43 to 63 per cent) to which the importance of marketing employment varies in the larger Danish towns. There are thus strong spatial differences in the intensity of operation of the marketing system; spatial differences in the processes and components of the system are discussed later in this chapter.

III Marketing theory and the geographer's contribution

The geographer's contribution to the theory of marketing systems is of two types, the first being the extension of marketing theories to include the spatial dimension, and the second the indication of how geographical theories may be of value to the marketing scientist. It is beyond the scale of this essay to discuss in full the many theories of marketing (Schwartz, 1963) and the way the geographer's work may be applied to them. Only three marketing theories can be considered: the wheel of retailing, the product-life cycle theory, and the character-of-goods theory. In addition, however, a brief discussion of central place theory is included as representative of geographical theories relevant to marketing. These four are not theories in the scientific meaning of the word—in fact, in most instances they are barely hypotheses. Each provides, however, a conceptual model which may be tested against reality. Furthermore, in each case the theories are related (most usually) to consumer markets and are strongly focused on the retail trades, since other markets and distributive channels do not appear to have attracted the theorists to anything like the same extent. Apart from the important contribution of Vance (1970), the wholesale trades, for example, have received scant theoretical and empirical attention from geographers, whilst in marketing the work of Beckman, Engle and Buzzell (1959) and Bartels (1963) stand out in a list of studies of this topic.

1   The wheel of retailing

The hypothesis of the wheel of retailing concerns the process of retail development. McNair (1958) suggested that new types of retailer usually enter the market as low-status, low-margin, low-price operators and gradually acquire more elaborate outlets which incur higher investment levels and higher operating costs. These retailers eventually reach the position of high-cost and high-price outlets vulnerable to new low-price competitors. This circular effect is called the wheel of retailing. Studies of retail structure within urban areas suggest that, as the wheel turns so locations of the shops change from less convenient low-rent sites to more convenient and consequently more expensive sites. The circular process, furthermore, is likely to affect different regions and sub-regions at different rates, and consequently the wheel will both turn at different rates and also be at different stages in different regions. The circular process is largely a response to increases in living standards; consequently differentials in levels of affluence will provide opportunities for low-cost retailers to enter the market at the lower level. The receptiveness of areas to the innovation of low-margin retail types would thus be expected to be related inversely with the standard of living in an area.

The simple hypothesis of McNair is not totally acceptable. Examples of the developing retail structure in underdeveloped countries have shown that new retail outlets such as supermarkets are attractive to the higher-income consumer and are consequently at the upper end of the price continuum. Furthermore, in developed countries the re-planning of shopping centres often means that the new types of retail trading are attracted to new developments, although costs may be higher than in nearby older shopping centres. The very wide range of selling policies within any one retail method means that the generalizations implicit in the wheel hypothesis need not be applicable to all firms.

An adaptation of the wheel concept has been developed by Agergård, Olsen and Allpass (1968) who suggest a spiral process.

At certain times, depending on the existing retail structure and current standard of living, new types of shops are established which fulfil the same function in the retail and urban structure as a previous shop type. The new shops, however, function on a higher level, with regard to prices, quality and location. The *wheel* theory is thus placed against the background of a rising standard of living and a *spiral* is produced.

2 Product life-cycle theory

In simple terms this theory states that there are up to five discernible periods in a product's life time.

(i) *Introductory.* This may be a long period during which sales volumes increase only slightly. Sales levels are likely to fluctuate as test purchases are made. Some products will not go beyond this period.

(ii) *Growth.* Sales volumes begin to rise steeply together with profit levels. In this and the succeeding stage new uses must be found for the product by research and sales promotion.

(iii) *Maturity.* Profit levels reach a peak and begin to decline whilst sales volumes continue to rise. Towards the end of this period new competitive products may begin to appear.

(iv) *Saturation.* Sales volumes reach a plateau. If no alternative product is available the sales volume may increase, but only as fast as growth in the general economy. No new market is captured. Profit levels continue to decline.

(v) *Decline.* Sales volume decreases as market penetration is lost.

It is possible to interrupt or to rejuvenate this cycle at any point by increasing promotion and product research; but the costs involved would have to be offset against the cost and return from developing a new product.

The marrying of spatial diffusion theory to this concept of a product life-cycle offers a potential broadening of both theories. The work of Hägerstrand (1952), Rogers (1962), and others on the ways consumer information spreads through space is valuable to the marketing scientist preparing a product development plan. A product will be at different stages of its life-cycle at different points in space, whilst over a whole region it may be possible to see each stage of the product in a different area.

In the introductory stage sales will be concentrated in and around the initial nodes of introduction. Growth period characteristics will be apparent first in the initial introductory nodes, whilst the characteristics of introduction will be seen still in areas some way from the nodes. The speed of diffusion of the characteristics of each period will depend on the human and physical attributes of the consumers' environments. As the life-cycle works through and also diffuses over space, there will be some areas in which the product is declining whilst in other areas profitability remains high and sales are buoyant. The pattern of product acceptance and hence a product's development plan has a major spatial component. When product life-cycle theory is used in product development then the geographer's work on spatial diffusion may be useful in solving some of the problems of product planning.

3   Central place theory

Christaller (1966), drawing together and formalizing earlier work on the relationship of settlement centres to their surrounding areas, elucidated a theory relating to the location, size and spacing of consumer-market activities in settlement centres. The marketing application of central place theory, as Christaller's work is now known, lies in its relation to the size and commodity mix of market centres. The theory does not relate to market structure other than in terms of the number and variety of goods distributed. There now exists a considerable body of literature on modified or extended versions of the theory, but as Scott (1970) points out, 'The central place model cannot be readily reconciled with our knowledge of retail location derived from economic theory and from the behaviour of both consumers and entrepreneurs' (p. 12). As with all the theory related to marketing, central place theory is not a complete theory for the geographer studying markets nor for the marketing scientist interested in the spatial characteristics of markets. None the less the theory is of relevance in the solution of some marketing problems.

A mix of market commodities has to be offered at central locations so that consumers can group and concentrate their purchases. The basic postulate of central place theory is that demand occurs at nodes and the size of demand is directly related to the *centrality* of the node. Berry (1967) explains centrality as the attractiveness to consumers of a group of retail and service functions.

> For differing activities centrality therefore has meaning at different scales; in any area a *variety* of *central places* will thus exist. Businessmen located in some will attract consumers on a frequent basis, but only over short distances. Other places will be able to provide a greater variety of goods to much wider areas. The clusters of activity in these places vary, along with the sizes of the urban places in which the markets locate. (p. 3)

One of the theoretical claims of central place theory is that within a region there are discrete groups (a hierarchy) of settlements, with each group containing markets with similar commodity/service mixes and with similar total demand. Market centres and services may be grouped into *orders or levels* of the hierarchy, such that higher order centres duplicate the functions of lower order centres and in addition distribute a distinctive collection of higher order services. Market area sizes are related to this hierarchical ordering, and lower order areas are nested within the trade areas of higher order centres. Such a view is somewhat similar to the marketing concept of convenience goods, shopping goods and speciality goods, and the classification of shopping centres based on these types of goods. Central place theory offers an extension of this idea and presents a much more detailed grouping of goods and centres.

A further area where central place theory is of concern to marketing scientists is the extent to which the theory has pervaded recent thinking on the planning of shopping centres. Szumeluk (1968) indicates the very wide range of urban and regional development plans whose retail component is based on the ideas of central place theory, especially the hierarchical concept of shopping-centre size. Although central place models provide a dangerously simplified view on which to base plans for the provision of retail land uses, none the less they have been used (Surrey County Planning Department, 1967). Marketing plans must take into account the available channels of distribution, and in order to market products through planned centres it is as well to appreciate the theoretical views on which are based the centres' relative sizes and locations. The retailing element of any marketing system is influenced, of necessity, by the conscious efforts of man to regularize retail land use.

4 Character of goods theory

Aspinwall (1962) departed from the basic concept of central place theory, with its division of goods into convenience, shopping and speciality goods, and arranged goods along a continuous scale based on their marketing characteristics. The implications of Aspinwall's work for studies of retailing in urban areas has not been appreciated fully. Introducing the spatial element into his hypothesis may help to make more general his theory of the ways different products are distributed optimally through different channels. Implicit in Aspinwall's hypothesis is the consideration of the way the mix of channel sales varies regionally and subregionally. Within the strategy of any firm which is marketing goods for the consumer it is clearly important to know the relative channel composition in different areas of a market. Studies have shown that in Western Europe there are marked regional variations in the proportion of sales passing through various channels of distribution. Some of these variations may well be related, as implied by Aspinwall, to individual regional processes as well as to historical conditions which were particular to stimulating growth of some channels.

The basis of Aspinwall's theory is the consideration of goods in terms of five marketing characteristics, directly related to the trade channels used in their distribution:

(i)   replacement rate at which goods are bought (inversely related to the other four characteristics)
(ii) gross margin available to meet channel costs
(iii) services required by consumers
(iv) time involved in consumption
(v)   searching time on the part of purchasers.

The lower the total of measurements on the five characteristics the longer the channel of distribution. The characteristics of a commodity are not static through its life cycle; consequently the optimal distributive system for a given demand may change. The commodity will become better known as its becomes established and the distribution channel may well lengthen; however, as suggested by studies of the spread of innovation (Ohlsson, 1971), it is unlikely that a product will become established to the same extent in all regions.

IV Spatial processes in the marketing system

The marketing system represents a component of the environment which man has shaped and now must manage. The very nature of a system makes exclusive classifications of processes invalid, for the processes themselves are interrelated. Five headings have been used in this section to allow a summary consideration of some of the spatial processes in the marketing system.

1   Consumer processes

One of the major goals of marketing is consumer satisfaction, whether the consumer is middleman or housewife. Geographical studies have concentrated particularly on the final consumer, with studies both of consumer demand over areas and of the spatial behaviour of individuals or groups of consumers. The delimitation of areas with populations having similar economic and social characteristics has been carried out for urban areas and for much more extensive tracts. The retail sales and general levels of consumer demand will be a reflection of these population regions. Marketing studies have shown that particular social and economic characteristics of the population correlate with particular demand patterns and purchasing patterns. The spatial relationships of the population characteristics will result in both spatial variations of demand types and spatial variations in the consumer characteristics of middlemen.

A second group of geographical studies focuses directly on the behavioural character of the final consumers, their taste preferences, and the ways they view the spatial gap between, say, their home and the retailer. This behaviourist view is well illustrated by a study in Iowa by Rushton, Golledge and Clark (1967). The problem they try to solve is 'to find order in the consumer's selection of his maximum grocery expenditure town, i.e. the town which receives more of his grocery dollars than any other town' (p. 391). This problem is one reflecting spatial behaviour characteristics.

The behavioural approach (see also Ch. 7) lately has incorporated studies of the demand-type approach in order to model the spatial behaviour of large populations. All these studies integrating the two approaches have been concerned with the final consumer. The spatial behaviour of, for example, the retail or wholesale firm has been ignored. The bald generalizations of central place theory, as they relate to the spatial behaviour of the marketing firm, do less than justice to a very complex topic which could provide geographers with a major area of applied research.

2 Processes in market channels

(a) *Studies of location.* Studies of retail, and to a lesser extent wholesale, location have been a particularly prominent area for geographical study. Research in urban geography has produced many case studies of retail locational requirements (Cohen and Applebaum, 1960; Applebaum, 1968) and processes (Getis and Getis, 1968) within urban areas. At the scale of the total urban unit, central place theory has been used frequently to attempt explanation of observed retail locational patterns. Studies based on rent theory and price theory applied to site selection, operating costs, and so on have been used at the micro-scale to model assumed regularities of spatial location in small shopping centres. As Donnison in a N.E.D.O. (1970) report has indicated, both approaches are useful but neither presents the whole picture. Theoretical and practical problems of retail location are matters for research by geographers. The intro-duction of geomorphological measurements of urban land quality into the economic rent theory models is one such area of research.

On a broader locational scale, the physical and economic geo-graphers are pooling their knowledge to increase the understanding of retail land uses. First, in the analysis of sites for out-of-town shopping centres the geomorphologist has much to offer on site selection. The construction of such large-scale developments may have considerable effect on the physical landscape. Such effects may be both short-term, during the physical construction, and long-term, resulting from persistent human tampering with landscape. Secondly, decentralization of retailing from established cities and the growth of suburban shopping centres provides an opportunity for a much more sympathetic planning of retail needs related to the physical landscape.

The different locational patterns and processes of retail channel forms, such as consumer co-operative, retailer co-operative and multiple trader, have been studied all too briefly in their spatial setting. None the less the available studies are important on the one hand to product planners wanting to gain access to as large a final consumer market as possible, and on the other hand to the urban and regional planners whose task it is to consider the requirements of the different channel forms. The intensity and direction of market penetration of a voluntary chain or multiple retailer, as seen from Scott's work (1970), is related strongly to a company's organiza-tion—for example, location of the firm's head office. The develop-ment of regional offices results in *geographical growth* of a different kind and is the landscape result of a more hierarchical management form.

(b) *Studies of structure.* Even at the broad inter-regional scale, major differences in channel structure may be seen. Fig. 6.3 shows some of these patterns for the French Planning Regions. If retail operations are divided into food and non-food trade then the scale, and other characteristics, of retail operations may be seen to vary markedly over the country. The role of the consumer co-operatives as a form of distribution also varies systematically through France, as in fact it does through all European countries.

In much the same way innovations and technological change in the retail channels have been analysed spatially—for example the acceptance of self-service trading in the grocery trade. In most areas of Great Britain self-service was adopted first by branches of the large urban consumer co-operative societies in the early 1950's.

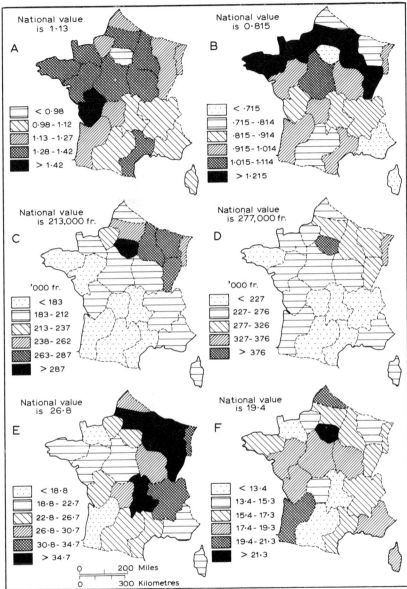

Fig. 6.3 Variations in retail structure in the French Planning Regions—1966. (A) ratio of numbers of food shops:non food shops; (B) ratio of investment in food shops:non food shops; (C) average sales per food shop; (D) average sales per non-food shop; (E) percentage of food sales passing through co-operative stores; (F) percentage of non-food sales passing through department stores. *Source:* French census of distribution, 1966.

For example, in South West England the Bristol Co-operative Society had over forty such shops by the mid 1950s. This was followed by a spread to rural societies, together with development by multiple traders in the towns. Thus in Bristol between 1959 and 1963, whilst the number of co-operative self-service shops remained little over forty, the number of independent self-service shops (mainly multiples) grew from eighteen to fifty-six. Thirdly, the voluntary retail chain (Spar, etc.) and private trader have taken up the innovation and also spread from town to country. By 1967, 90 per cent of independent self-service shops in the South West Region were affiliated to voluntary chains. Superimposed on this innovation by channel form is the spread of the technique into the regions. For example, the different channel phases of innovation of self-service in South West England may be discerned with, in addition, an overall spread of the technique from North East to South West. Thus in Gloucestershire rural private traders accepted the principle in about 1960 well before their counterparts in Cornwall, where it became established about 1965. The development of retail self-service and its wholesale counterpart of cash-and-carry indicates a clear spatially identifiable process in many European countries.

Channel structure, particularly as studied for the retail trades, also varies amongst towns, with groups of towns having similar processes defining their retail functioning. This situation exists in Northern Ireland (Dawson, 1972) where settlements may be classified, not on the basis of *size* of retail economy (as demanded by central place theory), but on the basis of *structure* of the retail channels. Although the full organization of these groups is not yet understood, their existence has implications for the marketing scientist because reactions to products would not necessarily be the same in towns of similar population size.

The description of market channels by geographers has a long tradition but researches on their economic, social and spatial processes and structures are at a relatively early stage. As yet these relate only to retail operation. Descriptions of other market channels (Bird, 1958; Best, 1970) have been carried out and provide a valuable basis for wider study of the spatial component of market channels.

3   Transport in the market system

A large section of the transport industry falls directly within the marketing system. It is not possible here to discuss the many transport-associated problems with which geographers have involved themselves. In trade area studies and traffic generation studies particularly, geographers have made a considerable contribution—and once again both relate most strongly to the retail trades with far less study of other middlemen.

(a) *Studies of trade areas.* The concept of a trade area has long been a foundation for studies in geography and marketing (Applebaum, 1968). The roles of freight rates and transfer costs are considerable in defining size and shape of market areas. Lösch indicates the importance of transport considerations, and many subsequent geographical studies have been based on Löschian arguments. By using different practical methods of definition and by case study, maps have been produced showing trade areas of a series of retail centres. Adaptation of Reiley's ratio method, in which trade areas are related to population of the centre and distance between competing centres, and the use of more sophisticated gravity models have been central to recent studies of the urban (retail) sphere of influence and the transport consumed by the shopper. Studies by Huff (1963), for example, introduce the probability principle into the definition of trade areas, with the ultimate construction of a mapped probability surface indicating the consumer's propensity to shop at a particular centre. To explain the size and shape of market areas, economic, social and physical measures have been used. The different size and shape of wholesale market areas in different physical regions in the U.S.A., for example, is a reflection both of different population characteristics of these regions and of the very different transport costs per mile in different physical landscapes. In many studies of market areas two major geographical variables have been isolated: transport cost and consumer dispersion. These two variables work alongside a whole series of other variables—spatial ones such as price differences and non-spatial ones such as the product on offer—to produce a complex, multivariable-defined trade area.

(b) *Traffic generation of different marketing land uses.* The generation and characteristics of traffic have been considered from the point of view of the numbers of vehicles arriving at and leaving from different land uses; but in addition much wider studies have been made; for example of accessibility for the routing of wholesale deliveries (Garrison *et al.*, 1959). The close connection of some of these geographical routing problems to methods in operations research is apparent from the work of Scott (1971). The relevance of such studies to warehouse location is very considerable. Wholesale warehouse location must consider a full range of traffic generation costs both for goods arriving from producers and for goods departing to retailers. The spatial problem to be solved is that of finding a location (or series of locations) which minimizes total operating costs for the wholesale function, but which does so within the constraints of retaining an acceptable level of service and without adversely affecting nearby land uses. Geographical techniques such as those of Scott for the analysis of complex spatial data have proved useful with similar problems.

Many of the transport studies indicate the considerable spatial interrelationships in the marketing system. Work by Wheeler (1960) in Scotland and Helle (1964) in Finland show how the travelling shop (a retail channel with a major transport cost) serves the needs of a dispersed rural community. In the Finnish case study it is pointed out how administrative constraints affect this particular retail channel and its transport costs. Another example of the interrelationships may be seen in trade-area studies where the consumer processes already discussed are an important factor in deciding what is a bearable level of transport cost. This level ultimately defines the trade area. The marketing scientist is often aware of all these relationships but is unable to analyse them and understand them to the extent that he would wish. To provide information on the marketing system and its relationships, market research has become itself a part of the operations of the marketing system.

4   Market research

The aspects of the market system discussed so far have been concerned with the physical distribution of goods; articles change ownership, or are transported, or are consumed. Market research, however, relates to information flow; the collection of data on the market system, its analysis and interpretation are important functions of market research. These studies of different market systems—comparative marketing—are of increasing relevance as international marketing grows in importance. The extension of a firm's operations into a new marketing environment means that the new market must be studied and entry into it planned. Thus market research is a response both to the complexities of a firm's marketing operation and to the considerable geographical variety of markets.

The study of the economic, social, and physical attributes—the geography—of a region, provides source material to the market researcher (Van der Zwan, 1967). As regional geography has become more analytical and has included more features of the regional environment, so its utility to marketing has increased. Many market research reports, for example those of Comart and Contimart, appear very similar to regional monographs. Their contents include discussion of size and form of the region, the customers (their location, social and economic characteristics), the channels of distribution, the nature of economic and environmental trends affecting the market, constraints to market operation as well as other geographical variables. There are differences in emphasis of course between the geographer's approach and that of the market researcher, but they are relatively minor ones. In both approaches it is the direct marketing environment which is of most importance. Thus the study of the structure of the distributive channels must include details of the major constituent firms, their past performance and future potential. Similarly, in respect of the constraints (negative and positive) to market functioning, the market researcher is interested in the overall picture and spatial variations in levels of property tax and sales tax, barriers to trade entry, the role of resale price maintenance, development grants, and other political and administrative influences affecting the marketing system. Often such material is collected by the geographer but is omitted from the final regional geography. As suggested in other chapters of this volume, the geographer by his very training is in a unique position in his interest in comparing environments. Much market research would appear to be a particular form of regional geography.

V Marketing and development issues

1   The case of the under-developed area

A theme found in several of the essays presented in this volume is the increased interest geographers have taken in the problems of economic development and social change in less developed regions. Marketing geographers have responded less than some other geographers, but none the less there is applied research in this field. In addition the remarkable growth in development economics has meant that marketing scientists have been considering the role of their discipline in economic development (Moyer and Hollander, 1968). In underdeveloped regions marketing, 'is in every one of these areas the least developed and most backward part of the economic system'. (Drucker, 1958, p. 253.)

It is in agricultural and food marketing that research has been most active. In many of the less developed countries up to 65 per cent of the consumer's budget may be spent on food. Frequently the distributive system is inefficient, with considerable over-employment.

The cities in under-developed countries present special problems for, as elsewhere, the cities are the major agencies of distribution but their rapid growth means that levels of economic and social development are scarcely keeping pace with increases in population. The marketing systems in these cities is an important component governing everyday levels of living. With others, Rostow (1964) has argued that for regional economic development a revolution in marketing must take place. Under such conditions even slight increases in distribution efficiency mean sizeable increases in real income and food availability to the consumer. Sale of farm surpluses in commercial markets means the use of intermediary middlemen. Efficient intermediaries, whether, for example, government agency or private wholesaler, stimulate both the production and ultimate distribution of goods. The spatial problems of development are epitomized in the problems of these intermediaries. Their essential purpose is the equation of production from areas of surplus to consumption in deficient areas under conditions of changing demand and supply. The solutions to this equation will differ from country to country and region to region (Prakash, 1962; Gould, 1963). In under-developed countries the increases in consumer satisfaction and agricultural production are seen by many as being very closely related to the development of an efficient middleman sector in the marketing system.

2   *The case of the developed area*

It is not just in the under-developed countries that more efficient marketing aids the development process. In the developed economy,

> with its growing abundance of goods, marketing takes precedence. Markets are dynamic. They are not assured to sellers. Consumers can choose from an expanding range of goods and services. They need not accept everything sellers offer. Rising incomes give them power to insist that producers develop new and better products to satisfy their expanding wants. (O.E.C.D., 1970, pp. 116–17)

This situation in the developed countries can be seen in respect of food production and marketing. Many of the developments in agriculture can be traced to a change in economic emphasis from production to marketing (Ch. 2). The increasing costs of marketing, which vary notably over a country, are in part responsible for this, but more important is a change in outlook by farmers. Agricultural development, its processes and patterns in economies where production is strongly consumer-orientated, is very closely related to the decisions of marketing scientists and levels of efficiency in the marketing system.

Development policies related to marketing and geography, are, however, not confined to the agricultural sector. One of the tenets of regional planning is the establishment of a regional marketing system, and thus marketing itself poses development problems. The development of a regional market system may involve the imposition of towns, which are the marketing agents, on virgin land as in the Polderlands of Holland (Thijsse, 1968), or the major revamping of an existing urban marketing framework, as planned in Ghana (Grove and Huszar, 1964) and India (Johnson, 1965), or—as indicated by French Regional Planning Policies—the use of growth poles to realign the present urban marketing pattern. In each instance there is a considerable geographical content to the regional plan, for the problems are those of allocating land uses at a regional level.

3   Planning marketing systems

Geographers interested in marketing in underdeveloped countries have studied the periodic markets which represent important market channels in these countries (Hodder, 1961; Good, 1970; Jackson, 1971). Studies of the size and periodicity of such markets could well prove useful for any plan to develop a wider range of marketing systems—as, for example, is envisaged in Tanzania, where a state co-operative channel is being introduced. Regional planning in most under-developed countries involves the selection of key centres which will act as diffusion centres for general economic growth and particular marketing reforms. Description and explanation of the relative marketing importance of the settlement centres in a region, together with the retail/wholesale structure of these centres, is vital to any attempt to produce a regional development plan.

The study of development and the ways marketing may be of value in the development processes have concerned economists far more than geographers. In underdeveloped countries the spatial problems of marketing reform provide problems to which the geographer can respond. The works of Best, Good and Gould indicate directions for study; but other approaches exist, for we know very little of the spatial variations in consumer attitudes, market channel structure, or local and regional constraints to marketing. Information on all these is basic to successful development planning.

The geographer's contribution to the management of marketing systems not only includes study and analysis of present market patterns but also must include prediction of future states of the market system. In general, geographers have been wary of committing themselves to forecasting the future. Prediction of the possible results of a series of out-of-town shopping centres in Great Britain, for example, is undoubtably a topic which fits closely with current research in marketing geography and with interest in general marketing. Such prediction is essentially controlled simulation. A second and very different type of forecasting which falls within the geographer's sphere of competence is prediction of *Delphi* type in which reasoned guesses are made of the type of invention and innovation which will affect the marketing system. An example could be the development of a few superstores in towns in the 10,000–40,000 population bracket, which would entail, among other things, massive changes in present management techniques. The geographer's interest in such changes and his utility in their planning would lie in the prediction of the spatial repercussions of particular innovations. The current lack of predictive geographical exercises relevant to marketing is probably due to the very existence of so wide a range of present-day marketing problems to which the geographer can competently apply his discipline.

References

AGERGARD, E., P. A. OLSEN and J. ALLPASS (1968): *The interaction between retailing and urban centre structure: a theory of spiral movement.* Lyngby.
APPLEBAUM, W. (1968): *Guide to store location research.* Reading, Mass.
ASPINWALL, L. V. (1962): The characteristics of goods theory *and* The parallel systems theory *in* W. LAZER and E. J. KELLEY, editors: *Managerial marketing: perspectives and viewpoints.* Homewood.
BARTELS, R., editor (1963): *Comparative marketing: wholesaling in fifteen countries.* Homewood.
BECKMAN, T. N., N. H. ENGLE and R. D. BUZZELL (1959): *Wholesaling.* New York.
BEST, A. C. G. (1970): General trading in Botswana 1890–1968. *Economic Geography* **46**, 598–611.
BERRY, B. J. L. (1967): *Geography of market centres and retail distribution.* Englewood Cliffs.
BIRD, J. (1958): Billingsgate: A central metropolitan market. *Geographical Journal* **124**, 464–75.
CHRISTALLER, W. (1966): *Central places in Southern Germany.* Translation of 1933 edition by C. W. BASKIN. Englewood Cliffs.
COHEN, S. B. and W. APPLEBAUM (1960): Evaluating store sites and determining store rents. *Economic Geography* **36**, 1–35.
DAVY, L. (1967): Le marché-gare de Nîmes-Saint Césaire. *Bulletin de la Société Languedocienne de Géographie* **90**, 169–214.
DAWSON, J. A. (1972): Retail structure in groups of towns. *Regional and Urban Economics* **2**, 25–65.
DRUCKER, P. F. (1958): Marketing and economic development. *Journal of Marketing* **22**, 252–9.
FISK, G. (1967): *Marketing systems.* New York.

GARRISON, W. L., B. J. L. BERRY, D. F. MARBLE, J. D. NYSTUEN and R. L. MORRILL (1959): *Studies of highway development and geographic change.* Seattle.

GETIS, A. and J. M. GETIS (1968): Retail store spatial affinities. *Urban Studies* **5**, 317–32.

GOLDMAN, M. I. (1963): *Soviet Marketing.* Glencoe.

GOOD, C. M. (1970): Rural markets and trade in East Africa. *University of Chicago, Department of Geography, Research Paper* **128.**

GOULD, P. R. (1963): Man against his environment: A game theoretic framework. *Annals of the Association of American Geographers* **53**, 290–97.

GROVE, D. and L. HUSZAR (1964): *The towns of Ghana.* Accra.

HAGERSTRAND, T. (1952): The propagation of innovation waves. *Lund series in Geography, Series B (Human Geography)* **4.**

HELLE, R. (1964): Retailing in Northern Finland: particularly by mobile shops. *Fennia* **91.**

HEWINGS, G. J. D. (1971): Regional input-output models in the U.K. *Regional Studies* **5**, 11–22.

HODDER, B. W. (1961): Rural periodic day markets in a part of Yorubaland. *Transactions of the Institute of British Geographers* **24**, 149–59.

HORTON, F., editor (1968): *Geographic studies of urban transportation and network analysis.* Evanston.

HUFF, D. L. (1963): A probabilistic analysis of shopping centre trade areas. *Land Economics* **39**, 81–9.

ISARD, W., T. W. LANGFORD Jr. and E. ROMANOFF (1967): *Preliminary input-output tables for the Philadelphia region.* Philadelphia.

JACKSON, R. T. (1971): Periodic Markets in Southern Ethiopia. *Transactions of the Institute of British Geographers* **53**, 31–42.

JOHNSON, E. A. J. (1965): *Market towns and spatial development in India.* New Delhi.

KANSKY, K. J. (1963): Structure of transport networks: relationships between network geometry and regional characteristics. *University of Chicago, Department of Geography, Research Paper* **84.**

LAZER, W. and E. J. KELLEY, editors (1962): Editorial postscript, *in Managerial marketing: perspectives and viewpoints.* Homewood.

MCNAIR, M. P. (1958): Significant trends and developments in the postwar period, *in* A. B. SMITH, editor: *Competitive distribution in a free, high-level economy and its implications for the university.* Pittsburgh.

MIERNYK, W. H. (1970): Long range forecasting with a regional input-output model, *in* H. W. RICHARDSON, editor: *Regional economics: a reader.* London.

MINTZ, S. W. (1956): The role of the middleman in the internal distribution system of a Caribbean peasant economy. *Human Organisation* **15**, 18–24.

MOYER, R. and S. C. HOLLANDER (1968): *Markets and marketing in developing economics.* Homewood.

NATIONAL ECONOMIC DEVELOPMENT OFFICE (1970): *Urban models in shopping studies.* London.

ORGANIZATION FOR ECONOMIC COOPERATION AND DEVELOPMENT (1970): *Food marketing and economic growth.* Paris.

OHLSSON, B. (1971): Service and spatial change, *in* A. G. WILSON, editor: *Urban and regional planning.* London.

POLANYI, K., C. M. ARENSBERG and H. W. PEARSON, editors (1957): *Trade and market in the early empires.* New York.

PRAKASH, O. (1962): *The theory and working of state corporations*. London.
ROSTOW, W. W. (1964): *View from the seventh floor*. New York.
ROGERS, E. M. (1962): *Diffusion of innovations*. New York.
RUSHTON, G., R. G. GOLLEDGE and W. A. V. CLARK (1967): Formulation and test of a normative model for the spatial allocation of grocery expenditures by a dispersed population. *Annals of the Association of American Geographers* **57**, 389–400.
SCHWARTZ, G. (1963): *The development of marketing theory*. Chicago.
SCOTT, A. J. (1971): *Combinatorial programming, spatial analysis and planning*. London.
SCOTT, P. (1970): *Geography and retailing*. London.
SURREY COUNTY PLANNING DEPARTMENT (1967): *Shopping potential in Surrey*. Guildford.
SZUMELUK, K. (1968): Central place theory: II: Its role in planning with particular reference to retailing. *Centre for Environmental Studies Working Paper* **9**.
THIJSSE, J. P. (1968): Second thought about a rural pattern for the future in the Netherlands. *Papers, Regional Science Association*, **20**, 69–75.
VANCE, J. E., Jr. (1970): *The merchant's world: the geography of wholesaling*. Englewood Cliffs.
WHEELER, P. T. (1960): Travelling vans and mobile shops in Sutherland. *Scottish Geographical Magazine* **76**, 147–55.
ZWAN, VAN DER (1967): Regional analysis as an instrument of marketing. *De Economist* **115**, 231–42.

7 Politics

J. N. H. Douglas

I Heritage

In 1963 the American National Academy of Sciences set up a Committee on Geography to review the status of the subject and to establish areas of study within the discipline which, because of their contemporary social significance, should be emphasized and supported as major research fields. In its Report (National Academy of Sciences—National Research Council, 1965) the committee concluded that there were three basic cultural processes at work altering the nature of the world in which man lives:

(i)   demographic increase and movement
(ii) technological development
(iii) political organization and administration.

Today, as government planning has largely replaced the laissez-faire approach in the attempt to direct environmental change, political organization and administration must be accepted as the most authoritative cultural process. Political geography, because it is concerned directly with the nature of political areas and the functioning of political organizations, therefore occupies a field of considerable practical significance. Growing interest in politico-geographic study, particularly in North America, during the last two decades reflects this contemporary significance and stems from the realization of the many ways in which the political process influences the development of human society. To approach social problems in the hope of both providing understanding and contributing to solutions, however, requires the existence of meaningful theoretical frameworks and well-tried methods of analysis.

Despite a distinguished heritage which stretches from Greek times to the period of the last hundred years, when geographers such as Ritter, Ratzel and Whittlesey were active in the analysis of political-environmental relationships, political geographers have only recently begun to establish a useful body of theory and method. The relegation of political geography in the twentieth century to a weak and peripheral position in most teaching programmes is, in fact, a result of the excesses of some of its former workers. Many early studies accepted too readily the deterministic link between the physical environment (particularly its climatic component) and the nature of man as a political being and affirmed it as a basis for political action. While the deterministic political writings of Aristotle (fourth century B.C.) and Strabo (first century A.D.) can be viewed in the context of their limited knowledge of diverse environments, the more detailed development of deterministic theories in the late nineteenth and early twentieth centuries have bequeathed a most unfortunate legacy and are less easily condoned. Further, in the work of Ratzel (1896) environmental determinism was overlaid with organic analogies absorbed from the biological sciences and used to explain the nature of political entities (states). Ratzel argued that because the state is analogous to the physical organism it must be born, grow and expand or decay and die. The state must grow as its people's culture develops, and growth will naturally proceed through the annexation of smaller states and the envelopment of politically valuable positions. The acceptance of such laws of state growth and further organic theories in Germany before World War II led to the applied pseudo-science of Geopolitik and the intellectual justification of territorial expansion in terms of life space (Bowman 1942). It is hardly surprising that these developments gave rise to a strong suspicion of politico-geographic study in the postwar period.

While distorted geopolitical views were being applied for subjective ends, the development of more objective theories and methods of analysis by political geographers who rejected such a distorted approach would have helped to ward off the unfortunate consequences. Few useful developments took place, however, and in general political geographers simply described the historical development of the state and then concentrated on its territorial endowments in terms of both human and natural resources. Such a philosophy lacked analytical strength and can best be described as encyclopaedic. This approach, which can be seen running through the works of Bowman (1922), Whittlesey (1938) and Weigert (1957), provided much useful factual information—Bowman's work was aimed at providing a basis for American government decision-makers—but it failed to clarify the nature of political-environmental processes. Indeed, political geographers had consistently avoided the study of political process, arguing that its analysis fell properly within the scope of political science (Jackson, 1958). It was with some justification, therefore, that Burghardt (1965), in a review paper, stated that 'political geography as a developed field of academic geography scarcely existed before World War II'. (p. 229)

The first steps in the construction of the theoretical framework now used by political geographers were taken by Hartshorne in 1950 when he placed emphasis upon the study of political processes. He stated that 'the central problems of political geography must be considered in terms of the functions of state areas and indeed the state or any politically organized area is the spatial consequence of political process' (p. 104). By introducing the concept of *centripetal* and *centrifugal* forces creating integration, conflict and disintegration within the political area, Hartshorne pointed unerringly to the constantly changing character of political entities and to the causes and consequences of change. S. B. Jones (1954) built upon Hartshorne's work by outlining a simple model which clarified the relationship between political decisions and political areas. The model expressed as

Political Idea—Decision—Movement—Field—Political Area

shows how ideas and decisions create movement and circulation, establishing a field of political significance which will affect the nature of the existing political area or will, in certain instances, give rise to a new political area. The flow from idea to area is essentially dynamic and creative: for example, a new town policy can lead to a government decision which, when implemented, will result in new land-use and transport patterns, which in turn result in changed fields of circulation, and all these changes affect the nature of the state area. However, ideas do not evolve in a vacuum but are always to some degree conditioned by the reverse flow from area to idea. Thus the patterns and characteristics of existing political areas provide the environmental framework within which ideas germinate and decisions are taken and put into operation. New town policies, therefore, usually result from and are an attempt to change an unsatisfactory situation of urban over-concentration. Jones's work outlines in rudimentary fashion a model of the political system, and this is particularly valuable as it clarifies the aim of the National Academy of Sciences Committee on Geography (1965) in stating that 'for political geographers the determination of a research problem within its full system context is an essential step'. (p. 41)

II The political system

Hartshorne and S. B. Jones provided the impetus which has moved political geographers towards a greater understanding of the political system and to its establishment as a general framework for research. When considered in greater detail, it can be seen that every political system has two inescapable general characteristics. These are:

(i) the political process by which it functions
(ii) the territory to which it is bound.

1 *The political process*

The political process can be defined as the succession of actions or operations which man conducts to establish or maintain a political system. While this definition helps clarify the situation by expressing the process in terms of human behaviour, important questions remain. Are all processes which have some significance for the functioning of the state or which are in any way influenced by the state to be designated political processes? If not, should predominantly cultural and economic processes be separated from more direct political processes and designated as non-political or, more accurately, para-political processes? This problem, one of functional classification, is vital to the development of the political system framework. Within every political entity there exists a constantly changing mass of organized and unorganized social interaction which could be perceived and understood if man were able to comprehend all social processes at a glance. Because political geographers cannot achieve omniscience, this mass of interaction must be simplified so that it can be perceived more clearly.

The question, familiar enough to geographers, is: where and on what basis should boundaries be drawn between processes which in reality merge and are indistinct? At this point it is not spatial boundaries but functional boundaries which are required. An employee at a factory while at work may exchange views with his fellow workers on a forthcoming election. In his work he is contributing to the economic system; in exchanging views on politics is he contributing to a political process? The answer given depends upon the criteria chosen to define the political process. The work of Easton (1965), a political scientist, provides a solution when he defines the political process as 'that process which makes authoritative and binding allocations within the state' (p. 60). Political processes are, therefore, those which result from government decisions and which are made authoritative and binding by the jural law of the state. Thus the first major component of the political system is the government with its powers of decision. All other social processes, which by this definition are non-authoritative, can now be designated as forming the environment—the second component of the political system—within which political processes operate. The employee of the factory who exchanges political opinions while at work is therefore operating within the environment of the political system. A model of the political system can now be constructed showing both its environmental and governmental components and its operational processes (Fig. 7.1).

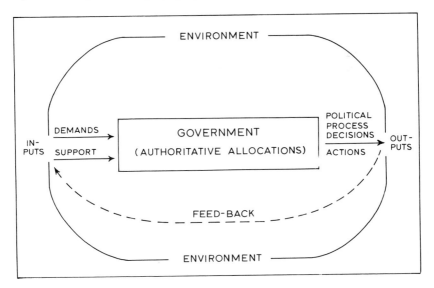

Fig. 7.1

The model shows that the relationships between the components of the system are not random or unorganized. As an open system it has the ability of self-regulation; the governmental component can respond and adjust to influences and stress from the environment. The influences from the environment resulting from para-political processes are of two kinds, namely demands and support (Fig. 7.1). These two inputs are interdependent as, to maintain the input of support, the system must respond to demands and attempt to satisfy them. The response comes through government decision-making and the resulting political processes. The political processes, in their operation, create changes within the environment. With this adjustment there is a feedback or information flow to the actors in the environment who consider the response and its effect and this may result in an increased level of support or generate further demands.

2   The political territory

Thus far understanding of the political system has resulted from the functional classification of social processes; further insights can be gained by considering its spatial characteristics. Political processes can operate freely and respond to demands only within a closed territory. The political boundary as the limit of sovereignty is also the limit of legal control by the political system. Any attempt to enforce decisions beyond its bounded territory means that the government is moving beyond its area of control, and its authoritative ability will be rapidly challenged by the governments of other political systems. In response to environmental demands the government must operate within a bounded area. The processes of the environmental component, however, are not so rigidly bounded. Economic, cultural and ideological processes are not wholly contained by political boundaries; their spatial characteristics are much more flexible and advance and contract much more easily. While it is true, therefore, that the most significant part of the environmental component of any political system is enclosed within its political boundary, demands can move across national boundaries and come from what, in the state context, is an international environment. Fig. 7.2 provides a theoretical example of the problems that result from the rigid delimitation of political space. The lack of success of the American government's rigid barrier philosophy and strict containment policy with regard to communism is an example which has drawn criticism from Cohen (1963), who shows the futility of attempting to enclose ideologies within watertight political boundaries.

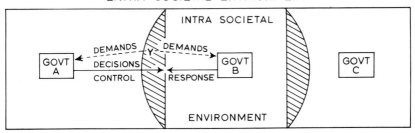

Fig. 7.2 The demands which minority group Y make upon political system B cannot be fully met because of the limit of control imposed by the political boundary separating political system B from political system A. Opposition by political system A to the demands of Y hinders legal action by political system B. The opposition by political system A may be supported by political system C which has a similar minority problem. Inability to respond fully to the demands of minority group Y may in turn create stress within political system B.

This model of the political system, which with modifications can be applied to the intra-national and supra-national levels as well as to the nation state level, establishes clearly the close interdependence of government-environment relationships. In common with most theoretical frameworks it can be criticized because its abstract nature creates problems of empirical application. This criticism cannot be refuted entirely, for new work within the social sciences has so far tended to concentrate on the elaboration of the theory of political systems rather than on the problems of empirical application (Young, 1968). While such criticisms cannot be overlooked, the political system model, like S. B. Jones's unified field theory, has the value of providing a logical housing place for more limited hypotheses and theories and it enables the researcher to place his work in a wider context and consider a greater range of stimulus and response factors. For the political geographer the model immediately places the environmental component in a meaningful functional and spatial context and clarifies its nature as a complex of dynamic processes.

III Major fields of politico-geographic study

As geographers as a whole, by tradition and interest, concentrate attention on environmental analysis, so political geographers can most fruitfully apply their abilities to the consideration of the environmental component of the political system. Three fields of study stand out as having considerable practical significance. These are:

(i) studies of political viability in terms of integrative and disintegrative forces at work within the political system
(ii) studies of political boundaries
(iii) studies of the behavioural process.

In practical terms the information and insights gained from these studies can be used to aid policy formulation in the attempt to satisfy environmental demands or to suggest changes in existing policies which have been shown to be ineffective.

1   Studies of political viability

Traditionally, political geographers have assessed the viability of political units by analysis of the distribution and homogeneity of cultural attributes such as language, religion, ethnic kinship, common cultural traditions and shared historical experience. The even spread of such attributes throughout the political area was taken as proof of political integration and, at the state level, of the presence of a national *raison d'être*. The existence of separate cultural attributes, whether concentrated spatially or dispersed throughout the political area, denoted the presence of minority groups which, it was generally accepted, created disintegrative forces and adversely affected political viability. While the compilation and cartographic representation of cultural attributes provides background environmental information, which can be valuable in plans to delegate power and locate internal state boundaries, the existence of viable nation states with diverse cultural characteristics precludes the acceptance of direct correlations between cultural homogeneity and political viability.

Cultural attributes are the bricks with which political viability can be constructed; the cement that binds a people together is found in social interaction and the formative political ideas that result. Deutsch (1953) established the importance of social interaction as a political factor and in developing his *theory of social communication* he showed how the study of cultural *composition specifications* could be replaced by *tests of performance*. Any community or nation is held together by the transmission, recall and storage of wide ranges of information. The transmission process is made possible by the communication equipment which exists within a community and is composed of learned memories, symbols and habits. Because social and political integration grow by this process of communication, the greater the density of communication flows the stronger the potential for integration. Cultural attributes, particularly language, can therefore be seen in a new light as factors conditioning the depth, spread and speed of communication flows. Deutsch's work leads logically to the analysis of transaction and interaction flows (see also Ch. 6, section I.2) as a means of measuring the strength of integration both within and between communities at all levels in the political hierarchy.

At the supra-national level Feldstein's study (1967) of transnational labour flows within the European Economic Community draws many useful conclusions on the relative strengths of national and international integration. The relative decline of international labour circulation and the consistent movement of immigrant workers back to their country of origin shows the continuing strength of national *raison d'être* and the weakness of a European identity. In practical terms, the overwhelming preference that the worker has for his country of birth suggests that it remains more realistic to formulate policies which take work to the labour supply rather than the labour supply to the work (see also Ch. 5). Also, the relative weakness of a European identity as shown by labour circulation reinforces the realism of the E.E.C. policy of pushing political goals into the long-term future.

A similar type of analysis by Soja (1968) based on telephone interaction in East Africa between 1961 and 1965 reflects how the prospects of federation diminished as interaction between Kenya, Tanzania and Uganda decreased in relation to the rapid growth in intra-state telephone interaction. Soja also shows that within each state there exist distinct areas which are held together by strong interaction; for example, within Kenya interaction is split significantly into two networks which parallel internal ethnic divisions of political significance. Finally, Soja shows that within each of the states there are areas which have few telephone linkages and so remain largely outside the integrative process set in motion by telephone communication. The information provided by the analysis of transaction flows has obvious significance for policy formulation and, as a study of the historical development of the Canadian transportation system by Wolfe (1962) shows, politicians have long been aware of the importance of social communication as an integrative factor and have attempted to control it through the planned development of transportation networks.

At the opposite end of the political scale, for example at the intra-urban level, analysis of transaction flows can be of value. Boal (1969) analysed local transaction flows within Belfast and used the resulting patterns to establish the location of sectarian boundaries enclosing religious ghettoes. The boundaries, not immediately obvious to the outsider, dominate community life and pinpoint zones of environmental stress. Such information can be used to establish short-term policies aimed at limiting civil disorder and to help formulate long-term urban planning policies which will minimize the potential for community unrest. However, as Boal points out, well-intentioned housing policies alone cannot remove residential segregation. Ghettos result from and reflect the polarization of human perception, attitudes and desires, and attention therefore must also be paid to the pre-decision behavioural process within which lie the sources of conflict and environmental stress.

The limitations of assessing the level of political integration by means of transaction-flow analysis must not be overlooked. The density of telephone, mail, trade and diplomatic transactions reflect only the potential for social communication; the content of the information accompanying the transaction, the way in which the message is sent and how it is received are of even greater significance. Without content analysis, which may prove impossible in many studies, the nature of the relationship between social transactions and political integration will remain unclear and consequently policy inferences and conclusions, while still of value, must be drawn with caution and seen in proper perspective. Feldstein (1967) shows how the status and behaviour of migrant workers can adversely affect the image perception of local workers and so result in a strong reaction against integration. Political geographers in the study of transaction flows are merely scratching the surface of the integrative process and much remains to be learnt, particularly from political scientists such as Jacob and Toscano (1964) who have developed a classification of integrative factors in which transaction-flow analysis is set in its proper context.

Finally, studies of transaction and interaction within the environment of the political system not only reflect patterns of integration but also forcibly point to zones of separation and areas which are poorly integrated within the system. These areas may house communities who actively reject integration and demand the spatial reorganization of political systems through the partitioning of territory or the relocation of political boundaries.

2   Studies of political boundaries

Political boundaries result from man's sense of territoriality and his desire to delimit rigidly an area within which he can preserve his cultural identity and mould his political future. Thus at the national level political boundaries demarcate the limits of sovereignty, jurisdiction and power of political systems. At the intra-national level, provincial and local government boundaries result from the internal spatial structuring of the state for the purpose of domestic public administration. Approaches to boundary study have evolved from early attempts at classification to the present emphasis on the functional significance of boundaries (Minghi, 1963). Today, political boundaries are seen to be of practical significance because of the barrier function they perform—separating political systems, controlling the spatial character of political activity, and influencing life and landscapes in zones adjacent to them. In turn the strength of the boundary's barrier function reflects the state of relations between the socio-political communities it separates. Political boundaries which are based on mutual agreement or which have gained acceptance over time will usually have a much less significant barrier effect than boundaries resulting from disagreement and partition.

The utilitarian motive behind boundary studies shows in the early works of Holdich (1916) and Fawcett (1918) who argued the case for good boundaries and established, as they saw them, the principles on which boundaries should be delimited so as to safeguard international peace. Such work helped in the acceptance of the anthropo-geographic principle which, after World War I, located political boundaries in relation to cultural patterns and human desires rather than to strategic factors or notable physical features. The limitations of the anthropo-geographic principle were also established at an early date by the geographical work of Brigham (1919) who showed how, in the European context, the acceptance of national feelings as a basis for boundary delimitation inevitably produced small states which, while culturally homogeneous, were weak in economic terms. In highlighting the conflict which arises between the anthropo-geographic principle and the need for economically viable political units, Brigham's analysis was indeed prophetic.

(a) *Boundary change and the barrier effects of boundaries.* The geographical approach, being concerned with the spatial aspects of environmental processes, is particularly valuable in showing the way in which boundary change and partition interrupts established patterns of circulation and causes disruption of human life. Hartshorne's classic study of the effects of boundary change in Upper Silesia established a methodological approach which remains valuable to the present day (Hartshorne, 1933). By mapping, in this densely-populated urban area, the distribution of industrial sites, considering accessibility to transport lines, power and water supplies, and by superimposing on them patterns and densities of population movements, for example journey-to-work movements, he was able to show clearly the adverse economic and social effects of boundary change. A study by E. Jones (1960) of the Northern Ireland/Eire boundary in the post-partition era shows how the superimposition of a political boundary causes refugee movements and population reorientation. The analysis of population census data on religious denomination for the period 1911–51 establishes that after partition in 1920 the Protestant population in the boundary zone area within Eire drastically declined, causing many human problems not least of which was the disruption of the economic infrastructure and urban decay.

In practical terms the entrenched character of national boundaries inevitably prevents the implementation of logical boundary relocation policies suggested by geographic analysis. Yet, while illogical boundaries cannot be changed, the geographic analysis still has the practical value of establishing the boundary's disruptive effect, and it can be used to aid the formulation of policies which will alleviate the hardships experienced by the populations living in boundary zones. The study of boundaries which cut through areas of important natural resources and prevent integrated economic development is also of practical value. Boundaries which cut across and create international rivers give rise to problems of water-resource use for both upstream and downstream states. Studies of the Columbia River Treaty by Sewell (1966), the disputed waters of the Jordan by Smith (1966) and the Indus by Karan (1961) show how difficulties arise regarding water rights and how comprehensive planning strategies are seldom implemented when more than one political system is involved (see also Ch. 3). While the disputed use of natural resources can cause political conflict, studies that show the economic disadvantage of divisive boundaries can be used to strengthen arguments in support of international agreements and co-operation. As Held (1951) shows in an excellent regional study of the Saarland, which is German in language and sympathy but tied economically to France by its coal and steel industries, there are considerable advantages to be gained through functional integration, as suggested by the Schumann Plan which led to the development of the European Coal and Steel Community in 1952.

Mackay (1958) developed a new approach to the study of the barrier functions of boundaries by analysing the interaction (based on long-distance telephone calls) between selected Quebec cities, Quebec and Ontario cities and Quebec and American cities. By first determining the expected interaction, by the formula $P_A P_B/D$ where P_A and P_B are the populations of the two interacting cities and D is the distance between them and secondly plotting these theoretical values against the actual values, Mackay showed how the level of interaction between Quebec and Ontario cities was only 20 to 10 per cent of the expected interaction between correspondingly sized cities within Quebec, while that between Quebec and American cities fell to 2 per cent. This, Mackay suggested, may be interpreted as a measure of the influence of the boundaries. Thus Ontario cities behave as if they were five to ten times as far away as Quebec cities of the same size and separation and those in America as if they were fifty times as distant. While Mackay's conclusions can be criticized on the basis of the interactance formula, he was nevertheless able to express the effect of the boundary in clear quantitative terms.

(b) *Boundaries as factors of location.* By showing how the Northern Ireland/Eire boundary adversely affected urban service centres, E. Jones (1960) was exemplifying the practical importance of political boundaries as economic location factors. Jones's analysis, in effect, adds another case study supporting the original work of Lösch (1954) who established the ways in which a political boundary, by interrupting free economic circulation and focusing price differentials, gives rise to advantageous and disadvantageous economic locations. Taking the United States/Canadian boundary as an example, Lösch compared retail sales per capita in Windsor, Ontario, near the boundary opposite Detroit, with those of London, Ontario, which is very similar in size to Windsor but further away from the boundary. The comparison showed how in Windsor the sales volume of goods which were cheaper in the U.S.A. was very much lower than the sales volume of the same type of goods in London; for example, tobacco sales per capita in Windsor were only 65 per cent of those in London. Conversely, sales of goods which were cheaper in Canada were significantly higher in Windsor than in London; for example, fur sales in Windsor were 335 per cent of those in London. An early study by Ullman (1939) showed how state boundaries within the United States acted as factors of location and in time gave rise to distinct landscape differences in adjacent boundary zones. Thus his systematic analysis of the Rhode Island/Massachusetts boundary shows how the pattern of industrial sites can be explained in terms of differential state tax structures.

Recognition of the way in which boundaries act as barriers and thus as factors of location has direct practical significance in the successful siting of industrial and commercial enterprises and for politicians attempting to attract investment and implement effective industrial development policies.

(c) *Internal state administrative boundaries.* The importance of intra-state boundaries as factors which influence the efficient administration of public services has attracted the attention of political geographers. In the United Kingdom the problem of devising a new local government structure which will be acceptable to the majority of the population and permit efficient public administration has been the subject for several government commissions since 1945. A solution to the problem has not been brought nearer by the general acceptance that the existing pattern of areas and boundaries is out of touch with present-day realities. Environmental change, as expressed in urban growth and the expanding range of circulation, contrasts strongly with the rigid pattern of local government which, in the U.K., has remained largely undisturbed since the nineteenth century. Downs (1961) pointed to the particularly serious effects which outmoded local boundaries have on urban and metropolitan government. His conclusions on the spatial, financial and planning problems that arise were drawn from a consideration of the North American environment, but they apply with equal force to the U.K. situation.

The model of the political system (see Section II), provides a useful geographical approach to the problem. It establishes the reasons why the local government pattern should be fitted spatially to the patterns of circulation within the environment of the system. A close correlation of political and environmental space will reduce to a minimum the disruption to everyday life which usually results from political reorganization, and will provide a basis for integrated administration and co-ordinated planning. The lack of such spatial correlation, as shown by the present local government system in the U.K., is the basic cause of environmental dissatisfaction and administrative inefficiency (Robson, 1966). Geographers have made a significant contribution to this political problem by providing insights into the nature of environmental circulation. Central place and urban field studies have established the overwhelming significance of nodality and shown how circulation creates functional regions which integrate central city, suburban and rural components (Davies, 1966). The application of these findings to local government reform plans is shown in the work of Green (1959), who analysed transport and population circulation patterns in the South East Lancashire conurbation, and by Douglas (1968) who delimited local circulation fields based on work, retail, educational and recreation movements in South Yorkshire and used the information to clarify spatial anachronisms in the existing pattern and suggest a plan for reform. Minghi (1960), in comparing critically the political fragmentation of the metropolitan areas of Washington, D.C., and St. Louis with the area-wide requirements of water supply, sewage disposal, transportation and planning bodies, shows how largely similar approaches have been used profitably in the American context.

The emphasis placed upon the city (functional) region principle in the Report of the Royal Commission on Local Government in England (Redcliffe-Maud, 1969) is a clear reflection of the practical contribution made by geographers in the U.K. to the understanding and solving of the spatial problems of public administration.

3 Studies of the behavioural process

As the preceding sections show, geographic analysis of political integration and political boundaries concentrates on the nature of human behaviour in space. The transaction flows which reflect the potential for social communication, and the characteristics of environmental circulation which show the barrier effects of boundaries, can be located by geographical co-ordinates before the resulting patterns are analysed. While such studies have much practical significance, geographers have contributed less to the understanding of the motivation behind human activity. The fact that men engage in particular behaviour patterns and make one decision from among a number of alternatives poses problems which, until recently, the majority of geographers have tended to avoid. The importance of the pre-decision stage, namely the complex mental process by which man responds to environmental stimuli, has however been noted in geographic studies. Minghi (1963), in a review of boundary studies, notes how the establishing of boundaries and the functions they perform reflects the values and the attitudes which motivate the groups they separate. Boal (1969), in the study of residential segregation already outlined, points out how plans to prevent the evolution of ghettoes must concentrate on removing the sources of segregation.

Research on human motivation has come to the fore through the interdisciplinary schools of behavioural scientists who have attempted to understand the processes of perception, evaluation, attitude formation and learning which precede the decisions and actions of man. Accepting that the key to solving the most complex environmental problems lies hidden in the pre-decision stage, geographers have now turned their attention in particular to the study of environmental perception. As Kasperson and Minghi (1969) show, the spatial variable is a most significant element in the make-up of human perception and can influence human choice in many ways. From the innumerable quantities of processes and stimuli that compose the environment, man selects and constructs a simplified image of his environment This set of images or perceptual field is invariably distorted, to some degree, by his inherent values and prejudices. The practical significance of distorted images of locations, distances, distributions and places lies in the fact that man acts and reacts in accordance with his perceived world rather than in relation to the real world. Both Kirk (1951) and Sprout and Sprout (1956) showed the significance of the gap between the behavioural environment (i.e. the perceptual field) and the operational environment (i.e. the real world) in the way in which it affected both the formulation and the success of political policies. In a most dramatic example, Sprout and Sprout (1957) outline how, in the lead-up to the attack on Pearl Harbor, the American commanders remained totally ignorant of the approaching Japanese task force. Their behavioural environment contained no hostile force, and though that force was part of the operational environment, it was not related to American military decisions until the moment the attack began (see also Ch. 11, section I.4).

In less dramatic ways the works of Gould and White, Saarinen and Lowenthal show the value of geographic study of environmental perception. Gould and White (1968) analysed the perception in a sample of British school leavers of the residential desirability of different regions in Britain. After showing how such perceived environmental elements affect attitudes, they used the analysis to rank the attractiveness of different parts of Britain. As such information reflects the willingness of job-seekers to migrate to find employment, studies of this type have obvious significance for industrial location policies. Saarinen (1966), enquiring into the perception of drought hazard in the Great Plains, used a sample questionnaire approach to perception testing and compared the results with a quantitative study of drought. The results show how the farmer seriously underestimates the drought hazard and how this leads to a sense of false optimism and a consequent lethargy with regard to the implementation of farming practices which could preserve better the fragile ecological balance. Although not framed in terms of political consequences, this paper, together with similar work by Kates (1962) and Burton and Kates (1964), has significant implications in terms of environmental planning policies. Lowenthal (1958), considering the more overt political topic of the choice of capital city for the West Indies Federation (1958–62), shows how the differing perception of the people of Jamaica, Trinidad and Barbados, the three main island contenders, was preventing the choice of a generally acceptable location. Lowenthal considers the behavioural decision process by means of a close analysis of government reports, newspaper coverage and local literature, from which he is able to establish both the spread and the strength of inter-island perception and show its importance to the decision process. Other works by Inglehart (1967), using cross-national attitude surveys based on age groups, and Deutsch (1966), using content analysis of newspapers and interviews of elite groups, draw interesting though contrasting conclusions on perception of and attitudes to European political integration.

This account considers only the perceptual element of the behavioural process; more comprehensive coverage of the behavioural process in general and of perception studies in particular are found in the work of Kasperson and Minghi (1969) and Brookfield (1969). The account, however, does show that pre-decision motivation studies hold considerable potential for political geographers who wish to relate their work to contemporary political problems. Yet as man is a multi-goaled being who at any one time may pursue a variety of social, economic and political goals, some of which may be incompatible, geographers who wish to study the behavioural process may well achieve most within a framework of inter-disciplinary research.

IV Politics and environmental management

Studies referred to in the preceding sections show how geographers are concerned primarily with analysing and understanding environmental processes. The value of this work lies in the basis that it provides for suggesting and formulating effective public policies. In terms of the political system, geographic study can be seen to concentrate on the input or environmental demand stage which clarifies the nature of the environmental impress on politics. As a consideration of public policies frequently shows, however, environmental inputs do not necessarily bear a direct relationship to government outputs in terms of decisions and actions. A return to the political system model helps to resolve this problem. Demand inputs begin as a mass of conscious, but usually undifferentiated, wants or desires; because of their unsorted nature, they go through a long conversion process which trims down and classifies them into recognizable issues which can then be met by government response outputs. Therefore, although many demands are articulated, most fall by the wayside in the conversion process and only a restricted number reach the output stage. Because of the conversion process, geographers cannot hope to explain government policies through the analysis of environmental inputs alone; consideration must also be given to the output stage if the political impress on the environment is to be clarified. Geographers, due to their traditional concern with the environment, have seldom studied the output stage of the political system as a central theme, yet it is the dynamic processes resulting from political decisions which most strongly condition environmental change. This is particularly true in the present day, when the strength of forces contributing to air, water and land pollution is so great that governments alone have the power to supervise their actions and preserve the quality of the environment.

While not all geographers can be expected to shift their interest from the environmental input stage, political geographers could make a worthwhile contribution to the study of environmental change by showing more concern with the political impress on the environment and developing approaches to the study of environmental management policies (see also Ch. 11, section II). Cohen (1966) considered the government–environment link to be of such practical importance that he suggested geographers should concentrate on and develop what he called 'the geography of policy'. He argued that, while geographers have often studied the development and use of natural resources in terms of private entrepreneurial activities, not nearly enough interest has been shown in the vital role played by governments and other public institutions.

1   Framework for the study of public policy

Although the now classic work of Whittlesey (1935) outlined the ways in which politics mould the environment, it is only recently that political geographers such as Kasperson and Minghi (1969) have begun to develop general frameworks for the study of public policies. Kasperson and Minghi suggest that the analysis of any public policy can use with profit the following approach.

(a) *Consideration of the public goals of policy.* Each policy adopted by a government has a specific goal or set of goals which, it is hoped, the implementation of the policy will achieve through time. These specific goals are not always made explicit and policies are often justified because they will contribute to 'the greatest good for the greatest number' or because they will help achieve 'a balanced and prosperous environment'. While such general goals exist, every study must clearly establish the more limited aims of each policy under consideration whether the policy be concerned with human resources, for example employment or urban renewal policies, or with natural resources such as water and public open space. An understanding of the specific aims of policies is important because all policies are concerned basically with the re-allocation and redistribution of human and natural resources. Thus the funds upon which policy implementation depends will be drawn from all parts of the political system to finance projects in spatially concentrated zones, for example the central government funds that are used to implement regional development policies. While the aims of policies must be understood in terms of the projected spatial redistribution, it must not be forgotten that policies also make functional re-allocations. Policies concerned with the storage and provision of water must consider the demands of domestic and industrial users as well as its uses for power generation, fishing, recreation and even pollution abatement (Ch. 3). By establishing the specific aims of policy and the nature of intended spatial redistributions and functional re-allocation, insights are provided into the areas and the actors within the environment that will be most immediately involved. Study of policy can then be focused more keenly and possible problems and results considered.

(b) *Analysis of the decision-maker's evaluation and establishment of assumptions upon which policy rests.* Every policy which is formulated and put into operation refers in some degree to existing environmental situations. Thus, for example, urban renewal policies are a response to urban malaise and natural resource policies to the needs and demands of competing users. To begin to understand why certain policy lines are followed, one must ask 'How much and what sort of information does the policy maker have at his disposal, and what conclusions does he draw from it on which to base his assumptions regarding policy?' Consideration of reports of commissions set up by government and of studies submitted independently by interested bodies, together with the record of parliamentary debates in which the relevant issues are discussed, leads to an understanding of the decision-makers' information field. The record of parliamentary debates can also highlight the decision-maker's evaluation of the information provided. Limitations in his information and its effect on resulting evaluations and predictions can then be clarified. The attempt to assess the decision-maker's general motivation which colours his evaluation of information and therefore affects his assumptions provides another aspect of study. In most cases his motivation results from an economic rationality and cost-benefit approach to the problem at hand. While it may prove impossible to establish the decision-maker's motivation with complete confidence, a policy based on purely economic considerations can frequently have damaging limitations which will prevent successful implementation; for example local government reform plans based upon the principle of efficiency and value for money in local government services can run counter to the wishes of many inhabitants who have pride in and wish to maintain local political units which safeguard their local identity. As Kasperson and Minghi (1969) point out, 'Public managers of the environment must recognise not only a *political* economy but also that the "public good" is a protean beast' (p. 432).

(*c*) *Problems of policy implementation.* At the implementation stage, policies can give rise to conflict and encounter debilitating setbacks which will give rise to unforseen results and cause the policies to fail in terms of their original aims. The fragmentation of executive power between government departments who may wish to emphasize different aspects of policy decisions creates internal government conflict and prevents successful policy implementation. The Economic Planning Regions, set up in England in 1964 to carry through the ambitious aims of the Labour Government for economic growth, showed the problems of overcoming departmental spheres of interest. The main executive body in each region, known as the Regional Economic Planning Board, was composed of civil servants representing the interests of the main Government Departments in the region. This structure immediately established potential for conflict between overall regional and specialized departmental goals. Major problems also arise when separate policies being implemented by different government departments have adverse mutual effects. Transportation plans and urban development priorities frequently clash with rural preservation plans and policies for public recreation. Marts and Sewell (1960) exemplify the adverse repercussions that one policy can have on the other in their study of the conflict between fishing and the development of power resources in the Pacific Northwest. As the section on boundary studies shows, the political partitioning of space can seriously inhibit successful policy implementation. While the most obvious effects result from the existence of international boundaries, intra-state boundaries, as demonstrated by Quinn (1968) in a study of water transfers in the American West, can also create obstacles to efficient policy development and implementation.

Finally, environmental reaction to policy decisions can seriously affect the policy both before and during its implementation. Despite the desires of successive governments since 1945 to reorganize local government in England, reform proposals are only now approaching the implementation stage. The strength of vested interest stemming from the existing system has successfully criticized and opposed government action for over 25 years.

(*d*) *Environmental results of policy implementation.* As public policies are necessarily concerned with the re-allocation and redistribution of scarce resources, environmental change is a logical consequence of policy implementation. However, because policies which are concerned to achieve a desired future state of affairs are based on human evaluation and prediction, policy implementation can give rise to unforeseen as well as expected results. Indeed, the practical aim of policy study as outlined must be to establish the environmental results of policy and to understand and seek the causes of unforeseen change, whether it is due to faulty evaluation by decision-makers or to problems of implementation. As the input—conversion—output process within the political system is going on, the information gained from the study of policy can be put to practical use to suggest policy change and develop new strategies which will be more likely to achieve the desired environmental goals.

2   *The case of Craigavon*

This general framework for the study of public policies can now be applied briefly to a specific case. In 1965 the Northern Ireland Government adopted a policy of developing a new town based on the existing towns of Lurgan and Portadown, 50 km west of Belfast. The new town, to be called Craigavon, was planned to grow in population from 36,000 (1961) to 100,000 by 1981.

(*a*) *The aims of the policy.* The overall goal of the policy was to achieve a more balanced pattern of population and economic growth within the political area. The specific aims were to set up a new urban focus with a growing population in part attracted from the oversized primate city of Belfast (pop. 416,000 in 1961) and to create a competitive industrial base which could attract, with the aid of Government economic incentives, new industry from outside and within Northern Ireland. The aim of attracting population from Belfast was strengthened by a further policy decision to place a 'stop-line' on growth around the city.

(*b*) *The decision-maker's assumptions and evaluation.* The decision-maker's evaluation of the environmental situation was based largely upon two reports commissioned by the Government. The Matthew

Report (1963) emphasized how the overconcentration of population in greater Belfast (population 568,000 in 1961 and 40 per cent of Northern Ireland's total population) was creating an imbalance within Northern Ireland as a whole and causing problems of urban quality within Belfast. For example, since 1945 when the Government developed its industrial incentive policy, 60 per cent of industrial concerns attracted to Northern Ireland has set up within 40 km of the centre of Belfast. The Report then suggested the establishment of a new town and the 'stop-line' policy for Belfast. The Wilson Report (1965) on economic planning reinforced the new town strategy when it pointed out that modern industry requires relatively large and expanding centres of population. The centres it selected as most promising for future development were, with the exception of Londonderry, located in the east of the country and reflected the population imbalance. The assumptions were, therefore, that the Lurgan/Portadown area was the optimum location for the new city, that the existing population would provide the required base on which to generate industrial growth, that industrial incentives would attract industry to the new city and that with this industry and a well-planned urban environment, population would be attracted from Belfast.

(c) *Obstacles to successful implementation.* Despite the emphasis put on the attractions of Craigavon, industrial enterprises enticed to Northern Ireland by special grants often preferred to locate elsewhere; the perceived economic advantages of greater Belfast frequently remained the decisive location factor. In contrast to the planners' low consideration of Belfast's urban environment, the people of Belfast tended to perceive Craigavon as an embryonic city set in the backwoods. Further, the Government's road-building policy for a motorway linking Belfast with Craigavon to provide easy access to Belfast's docks meant that those people from Belfast who found jobs in Craigavon also found it easy to commute daily between the two cities. This fact indeed questions the assumption regarding the location of a new independent city so close to Belfast.

(d) *The results of policy implementation.* Although the policy has not reached the halfway point of its allotted time span, it is questionable if it will ever achieve its aim of establishing a new city independent of and competing on equal terms with Belfast. New industry in Craigavon has found a ready labour force from among local workers who have come onto the labour market as old established industries such as linen manufacturing decline. The policy indeed has had more striking and generally adverse effects in other parts of Northern Ireland. In Belfast the 'stop-line' on growth has succeeded in limiting building space and causing soaring house prices in the private sector. As the stop-line has been set aside in certain cases to allow new industrial enterprises to be located near Belfast, a significant degree of environmental dissatisfaction has arisen. Outside greater Belfast, particularly in the west and south of Northern Ireland, Craigavon appears to be viewed differently and the signs are that population from these areas is more than willing to move to the new city when job opportunities appear. Thus, the more successful the new town policy is, the more likely it is to attract population from rural areas distant from Belfast and so aggravate the population imbalance which it was designed to counteract.

The consequences of prolonged political unrest in Northern Ireland preclude hard and fast judgements on the new town policy; nevertheless, this brief study of policy shows that the original evaluation and assumptions contained important limitations, not least in the area of human perception. It is clear that when Northern Ireland returns to normal conditions, alternative strategies concerning industrial and population incentives will have to be developed if the aims of the new town policy are to be achieved.

Politico-geographic research suffers from a lack of manpower, despite the fact that within the general fields of political integration, boundary study, the behavioural process and policy analysis there exist an abundance of topics of practical significance to the complex and overcrowded world in which we live. With such a wealth of exciting and practical problems demanding attention, political geographers already engaged in research must, with the aid of new developments in the social sciences, advance their knowledge in theoretical and methodological as well as empirical terms, so that, attracted by the potential displayed, more geographers will turn their attention to those important processes that link politics and environment.

References

BOAL, F. W. (1969): Territoriality on the Shankill-Falls Divide, Belfast. *Irish Geography* **6**, 30–50.

BOWMAN, I. (1922): *The new world: problems in political geography.* London. (1942): Geography versus geopolitics. *Geographical Review* **32**, 646–58.

BRIGHAM, A. P. (1919): Principles in the determination of boundaries. *Geographical Review* **7**, 201–19.

BROOKFIELD, H. C. (1969): On the environment as perceived. *Progress in Geography* **1**, 51–80. London.

BURGHARDT, A. (1965): The dimensions of political geography: three recent texts. *Canadian Geographer* **9**, 229–33.

BURTON, I. and R. W. KATES (1964): The flood-plain and the seashore. *Geographical Review* **54**, 366–85.

COHEN, S. B. (1963): *Geography and politics in a divided world.* London.

(1966): Toward a geography of policy (guest editorial). *Economic Geography* **42**.

DAVIES, W. K. D. (1966): The ranking of service centres: a critical review. *Transactions of the Institute of British Geographers* **56**, 51–65.

DEUTSCH, K. W. (1953): *Nationalism and social communication.* Cambridge, Mass.

(1966): Integration and arms control in the European political environment. *American Political Science Review* **60**, 354–65.

DOUGLAS, J. N. H. (1968): *Political geography and administrative areas, in* C. A. FISHER, editor. *Essays in political geography.* London.

DOWNS, A. (1961): Metropolitan growth and future political problems. *Land Economics* **37**, 311–20.

EASTON, D. (1965): *A framework for political analysis.* London.

FAWCETT, C. B. (1918): *Frontiers: a study in political geography.* Oxford.

FELDSTEIN, H. S. (1967): Study of transaction and political integration: transnational labour flow within the E.E.C. *Journal of Common Market Studies* **6**, 24–55.

GREEN, L. P. (1959): *Provincial metropolis: the future of local government in south-east Lancashire.* London.

GOULD, P. R. and R. WHITE (1968): The geographical space preferences of British school leavers. *Regional Studies* **2**, 161–82.

HARTSHORNE, R. (1933): Geographic and political boundaries in Upper Silesia. *Annals of the Association of American Geographers* **23**, 195–228.

(1950): The functional approach in political geography. *Annals of the Association of American Geographers* **40**, 95–130.

HELD, C. C. (1951): The new Saarland. *Geographical Review* **41**, 590–605.

HOLDICH, T. H. (1916): *Political frontiers and boundary making.* London.

INGLEHART, R. (1967): An end to European integration? *American Political Science Review* **61**, 91–105.

JACKSON, W. A. D. (1958): Whither political geography? *Annals of the Association of American Geographers* **48**, 178–83.

JACOB, P. E. and J. V. TOSCANO, editors (1964): *The integration of political communities.* Philadelphia.

JONES, E. (1960): Problems of partition and segregation in Northern Ireland. *Journal of Conflict Resolution* **4**, 96–105.

JONES, S. B. (1954): A unified field theory of political geography. *Annals of the Association of American Geographers* **44**, 111–23.

KARAN, P. P. (1961). Dividing the water: a problem in political geography. *Professional Geographer* **13**, 6–10.

KASPERSON, R. E. and J. V. MINGHI, editors (1969): *The structure of political geography.* London.

KATES, R. W. (1962): Hazard and choice perception in flood-plain management. *University of Chicago, Department of Geography, Research Paper* **78**.

KIRK, W. (1951): Historical geography and the concept of the behavioural environment. *Indian Geographical Society, Silver Jubilee Edition,* 152–60.

LÖSCH, A. (1954): *The economics of location.* New Haven.

LOWENTHAL, D. (1958): The West Indies chooses a capital. *Geographical Review* **48**, 336-64.

MACKAY, J. R. (1958): The interactance hypothesis and boundaries in Canada. *Canadian Geographer* **11**, 1-8.

MARTS, M. and W. R. D. SEWELL (1960): Conflict between fish and power resources in the Pacific Northwest. *Annals of the Association of American Geographers* **50**, 42-50.

MATTHEW, SIR R. (1963): *Belfast regional survey and plan* 1962 (2 vols.). Belfast.

MINGHI, J. V. (1960): The spatial pattern of key functions in the Washington metropolitan area. *Unpublished paper, Washington Center for Metropolitan Studies*. Washington, D.C.

(1963): Boundary studies in political geography. *Annals of the Association of American Geographers* **53**, 407-28.

NATIONAL ACADEMY OF SCIENCES—NATIONAL RESEARCH COUNCIL (1965): *The science of geography*. Washington, D.C.

QUINN, F. (1968): Water transfers: must the American west be won again? *Geographical Review* **48**, 108-32.

RATZEL, F. (1896): The laws of the spatial growth of states. Translated by R. BOLIN, *in* KASPERSON, R. E. and J. V. MINGHI, editors: *The structure of political geography*. London.

REDCLIFFE-MAUD, LORD (1969): *Report of the Royal Commission on local government in England* (3 vols.). London.

ROBSON, W. A. (1966): *Local government in crisis*. London.

SAARINEN, T. F. (1966): Perception of drought hazard on the Great Plains. *University of Chicago, Department of Geography. Research Paper* **106**.

SEWELL, W. R. D. (1966): The Columbia River Treaty: some lessons and implications. *Canadian Geographer* **10**, 145-56.

SMITH, C. G. (1966): The disputed waters of the Jordan. *Transactions of the Institute of British Geographers* **40**, 111-28.

SOJA, E. W. (1968): Communications and territorial integration in East Africa. *The East Lakes Geographer* **4**, 39-57.

SPROUT, H. and M. SPROUT (1956): *Man-milieu relationship hypotheses in the context of international politics*. Princeton.

(1957): Environmental factors in the study of international politics. *Journal of Conflict Resolution* **1**, 309-28.

ULLMAN, E. L. (1939): The eastern Rhode Island-Massachusetts boundary zone. *Geographical Review* **29**, 291-302.

WEIGERT, H. W., editor (1957): *Principles of political geography*. New York.

WHITTLESEY, D. (1935): The impress of effective central authority upon the landscape. *Annals of the Association of American Geographers* **25**, 85-95.

(1939): *The earth and the state*. New York.

WILSON, T. (1965): *Economic development in Northern Ireland*. Belfast.

WOLFE, R. I. (1962): Transportation and politics: the example of Canada. *Annals of the Association of American Geographers* **52**, 176-90.

YOUNG, O. R. (1968): *Systems of political science*. New Jersey.

8 Air pollution

Peter Cox

The subject of pollution is inevitably emotive, for the use of the term frequently implies that there is an agency introducing into our everyday lives something harmful, unpleasant or even toxic. Pollution is not an absolute state of affairs but is relative to a particular human being or group of human beings. In the case of air pollution the term is generally used to apply to gases or to particles of matter found in the atmosphere as a result of man's negligence and which are in some way deleterious to his environment. In this sense unpolluted air is not synonymous with pure air since except in selected environments, for example over the Antarctic Continent, most air contains at least particles of soil or dust however low the concentration may be. The control of air pollution only becomes really necessary when it is harmful to man either directly (e.g. affecting his health, see Ch. 9) or indirectly through its effect on his biotic environment (e.g. damaging the plants he needs). Air pollution is of man's own creation and is affected by factors both in his physical and his social environs. This therefore places the study of many aspects of air pollution well within the traditional scope of both human and physical geography. More importantly, contributions have been made towards an understanding of the problem of air pollution by the application of geographical methods, and through an analysis of the relationships between air pollution and other aspects of the environment. This chapter reviews the contribution of geographical studies to the search for interdisciplinary solutions to problems of air pollution.

I The constituents of air pollution

1 *Principal pollutants*

The main substances recognized as being air pollutants occur in the form of particles, droplets or gases. Particles range in size from large deposits, which fall out of suspension in the atmosphere, to fine smoke particles. The latter are so small that they only scatter short wavelengths of light, giving the smoke a bluish appearance as in the case of tobacco smoke. Gases which are regarded as pollutants— and there are many more of these than can be mentioned here— are generally those which, at sufficient concentrations, are harmful to plants or animals. Such gases are sulphur dioxide, various hydrocarbons, carbon monoxide and nitrogen oxides. Some gases can undergo chemical changes in the atmosphere, though the exact nature of many such reactions is not yet fully understood. Sulphur trioxide in the atmosphere can react with water particles to produce droplets of sulphuric acid, and in addition to the possible harm caused by the particulate acid, other soluble products can be formed such as ammonium sulphate (on reaction with ammonia). Such reactions, when the reactants are in very small quantities, do not necessarily occur spontaneously and so the chemistry is complex. Many industrial processes produce gases, some of which would be severely toxic if allowed to escape. Under this heading Turner (1955) mentions silicon tetrafluoride, which is produced in various industrial processes, and hydrogen sulphide. The emission of toxic gases, however, in most such processes is prohibited by law in the United Kingdom.

Smoke is produced by the combustion of solid or liquid fuels and consists of unburnt or partly burnt particles of the combustive material. In recent years it has been a serious problem for domestic heating appliances where efficient and total combustion is often not practised. In industrial plant, inefficient combustion is economically unacceptable and this greatly reduces the smoke pollution from these sources. Turner (1955) found that per ton of fuel used domestic chimneys produced twice as much smoke as industrial chimneys.

Smoke production has come under serious consideration in many countries and legislative action has been taken introducing smoke abatement laws and control zones coupled with the widespread introduction of 'smokeless fuels', resulting in considerable inroads being made into the extent and concentration of smoke.

The most common gaseous pollutant in Britain is sulphur dioxide. The degree of its occurrence is due largely to the fact that most coals contain sulphur, so that on combustion the sulphur is oxidized and released as sulphur dioxide. Other fuels also contain sulphur, and so a similar result can come from many sources. A second gas, carbon monoxide, can reach dangerous concentrations in urban areas and particularly in streets where there is heavy traffic. The gas is produced as a result of the combustion of motor fuels. Since carbon monoxide is denser than air it tends to fall and so the highest concentrations are reached near to ground level.

Meetham (1964) mentions the problems found in Los Angeles where petroleum products are used widely for heating and power in addition to fuelling vehicles. As a result of climatic and topographic conditions (these are discussed more fully in section II.3) the pollutants produced from petroleum products build up and constitute a considerable danger to the everyday life of the community. When strong sunshine falls on air containing hydrocarbons (produced from petroleum products) and specific concentrations of nitrogen dioxide, evidence in California suggests that ozone is formed. Meetham notes that research in California indicates that the chemical action of ozone in the atmosphere is responsible for all the harmful effects of smog in Los Angeles.

Once the pollutant has been emitted (assuming it does not fall out of suspension in the atmosphere and so become lost) it is subject to the meteorological conditions at the time. Scorer (1968) examines, in summary, various meteorological effects—these will not be considered as a separate entity here, but some will be mentioned.

2   *Distribution of pollutant sources and emission*

In many studies the types of sources of pollution for an area have been mapped and areas of predominantly domestic or industrial land use have been noted. This, however, requires some knowledge of the relative emissions of pollutants from different sources before the mapped distributions become useful.

The study of air pollution in Great Britain by Craxford and Weatherley (1968) concerns the trends in smoke and sulphur dioxide emission in Britain over the period 1952 to 1967. In addition the study also considers regional data. The estimates used in this study were derived from detailed annual statistics on fuel consumption for different purposes. A steady decrease in overall emission of smoke for the period 1954 to 1967 was observed. This decrease was much influenced by the provisions of the 1956 Clean Air Act, partly as a result of which considerable changes to industrial appliances were made up until 1961. It is shown that the emission of smoke from domestic heating has fallen since 1957 under the influence of legally enforceable smoke control zones. However, domestic emission of smoke in the early 1970s is still considerably greater than industrial emission.

In the case of sulphur dioxide Craxford and Weatherley found no comparable fall in emission. Until 1960 there was in fact a steady rise owing to the replacement of coal in power and heat production by oils containing sulphur, with the highest overall emission being reached in 1963, from which time to 1967 it dropped by six per cent. When sulphur dioxide emission was examined by type of source it was found that domestic emission had been decreasing steadily over the period (a fall of 30 per cent from 1958 to 1967) with a small downward trend in industrial emission, so that the increase to 1963 was attributed by Craxford and Weatherley solely to emission from electricity generating stations.

Having examined Britain as a whole, Craxford and Weatherley went on to examine the situation in London, and also investigated pollutant concentrations, and emissions on a regional basis for 1967, dividing Britain into eleven regions. Smoke emission from both domestic and industrial sources was highest in the North West and lowest in Wales, in general being higher in the north of England and Scotland, per million inhabitants, than in the south. The emission of smoke from industrial sources for all regions was low, and Craxford and Weatherley attributed the higher domestic emission in the North and Midlands to the extravagant use of coal stemming, in mining regions, from the free issue of coal to miners. In Wales the coal is low volatile and so produces little smoke. The regional emission of sulphur dioxide was much more complex, with industrial emission in all areas being higher than domestic emission. The North again showed the highest figures for both domestic and industrial sources, while London had the lowest result for domestic emission per million inhabitants and Scotland the lowest industrial emission. The recorded mean concentrations of sulphur dioxide tended to follow the pattern of emission from domestic sources, as was the case for Britain as a whole during the period 1952 to 1967.

The analysis described above by Craxford and Weatherley illustrates how a regional study of pollutant emission and the preceding general study can help to highlight the broad areas and principal source types (domestic or industrial) where further abatement of pollution emission is necessary or desirable. A summary is provided in Craxford and Weatherley (1971), with some additional information and remarks. It is unlikely that comparable analyses could be performed on a local scale with any degree of accuracy, although Clifton, Grimoldby and Sharp (1969) constructed a domestic source index for smoke, for areas within the South-East Lancashire connurbation, based on an assessment of the percentage of an area in four categories of housing density. A sulphur dioxide source index was also produced based on an assessment of industrial power production potential in the areas (see section III). Garnett (1967) made an estimate, from fuel consumption, of the expected sulphur dioxide output from industrial plant in Sheffield, but in contrast to Craxford and Weatherley the information was furnished by individual firms so that a much finer mesh of information could be obtained.

II Influences of the landscape on patterns of air pollution

1 The effects of land form

Features of the landscape can have a considerable effect on the local climate in a particular area. For example, there is ample evidence of cold air sinking into deeply cut valleys leading to the development of pools of cold air in the valley bottom. Such a flow is described by Scorer (1968) and is known as a katabatic wind. This cold air is, by the very process by which it was formed, colder than the air above it and so no convection can take place. This situation often occurs towards evening or in the night and can result in pollutants being trapped in the colder air beneath a warmer air layer. When there is a layer where there is an increase in temperature upward through the air, or when conditions remain isothermal through the layer, it is said to constitute a temperature inversion. In a valley situation the inversion conditions are aggravated by the downward air flow into it and further dispersal of the pollutants laterally is prevented by the presence of the valley walls.

A team of geographers working under Professor Alice Garnett have for some time been engaged in an air pollution survey of Sheffield. This has been documented in a number of places (Garnett, 1967; 1970). The survey was designed to study the distribution of pollution by smoke and sulphur dioxide and to relate these to certain features in the city. It has been noted that characteristics of the city, particularly its site, influence these distributions, and of these the wide range in relief and the deeply incised valleys which radiate from the core of the city are especially noticeable. Natural ventilation is prevented by the presence of the valleys and cold stable air tends to drain into them. However, in the city centre and industrial areas in the Don Valley the increase in temperature as a result of the urban development may reverse this effect giving cold hill slopes and warmer valleys. During anticyclonic weather the inversion formed due to the subsidence of air, which is warmed as it falls and traps colder air beneath it, frequently drops in level. In Sheffield, this situation often resulted in dense fog below the inversion with no possible route of dispersal for the pollutants.

The effect of a temperature inversion associated with an anticyclone prevailing over the British Isles is illustrated in Fig. 8.1. This shows the results obtained at two sites in Sheffield between 1st and 5th February 1965 from measurements of smoke concentration, lapse rate index, and wind velocity. When the lapse index is positive the inversion conditions prevail. The strongest pulses of pollution often occur when the stable conditions are breaking down and the lapse rate index approaches zero, resulting in the bringing down of pollution from above; this is known as *fumigation*.

This illustrates just one way in which the physical environment can influence the patterns of pollution that are present, together with the fundamental influence of the human environment in the distribution of pollutants.

2   The effects of land use

Section I has shown how pollution emission differs between domestic and industrial sources; and since housing and industrial land uses often occur in blocks within cities, the land-use pattern is important in the distribution of atmospheric pollutants. Other significant smaller features can also be discerned: for example parks, by virtue of the vegetation in them, can act as filters to pollutants. Bach (1971) has established the contrasting effects of industrial and airport conditions in greater Cincinnati on the relationship between the vertical distribution of the optical properties of pollution and vertical temperature lapse conditions (Fig. 8.2). The aerosol is capable of reducing the solar radiation received and so is represented by the percentage attenuation of radiation for each layer; measurements were taken from a helicopter.

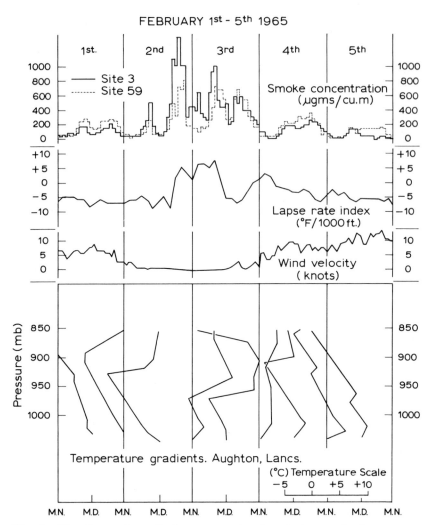

Fig. 8.1 Hourly smoke pollution at two sites in relation to the lapse rate index, wind velocity and temperature gradient (differences with altitude) February 1st–5th 1965. (The site numbers refer to recording sites used in the Sheffield Survey.) After Garnett (1967).

The study by Chandler (1965) of the urban climatology of London further illustrates some of the points made above. Especially low concentrations of smoke were found in the City of London itself, and this Chandler related to the small resident population, together with the effect of smoke-control regulations and the small number of smoke-producing industries. The importance of parks as filters was

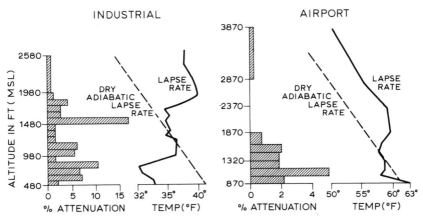

Fig. 8.2 Polluted layers and temperature lapse rate for two sites in Cincinnati. After Bach (1971).

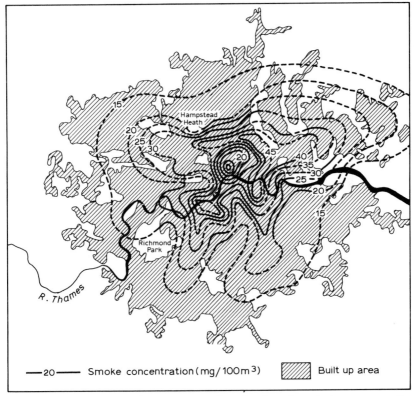

Fig. 8.3 Average smoke concentration in London (winter 1957–8). After Chandler (1965).

noticeable and important; this applied not only to smoke but also to heavier solid material. The effect is not simple and Chandler noted that when the concentration of smoke was generally high, then Hyde Park, for example, could be as smoky as the streets around. The filter effect was also present for sulphur dioxide, though it tended to vary very much with wind speed. Chandler's published maps, and particularly that for the winter months of 1957 to 1958 (summarized in Fig. 8.3), shows how the high-density domestic and industrial development of north-east London displayed high mean smoke concentrations, while central London remained much cleaner. The effect of even the wide valley in which London is situated may be evident in the relative cleanness which Chandler found in the up-standing tongue of land extending into London from Mitcham Common to Dulwich, and the areas of Richmond Park and Wimble-don Common, both of which are higher than the surrounding areas and which Chandler considers to be above London's pollution hood.

The influence of different types of land use can again be seen in the study by Clifton, Grimoldby and Sharp (1969) of levels of smoke and sulphur dioxide pollution during the winters 1961–8 in South-East Lancashire. They found that the recording sites in heavily industrialized areas such as Trafford Park or commercial areas such as central Manchester or Bolton suffered relatively little smoke, while in some of the suburban housing areas the concentrations of smoke recorded were higher than might have been expected.

3   The case of the Los Angeles Basin

So far, however, only the problems of smoke and sulphur dioxide have been mentioned, so that it may be of use to point out that environmental factors can influence other pollutants. The problem of pollution in the Los Angeles Basin has already been mentioned briefly. A summary account of the basic environmental factors can be found in the *Air Pollution Handbook* edited by Magill, Holden and Ackley (1956) (p. 38). On the whole this text tends to be specifically technical and is not recommended for general use; however, in this instance the summary is particularly concise. Los Angeles County consists of a basin fronting on to the Pacific Ocean but with mountains on all other sides. Warmed air from a semi-permanent high pressure system in the Central Pacific arrives in California much warmer than the surface air. The effect is to create an inversion layer which is dammed up against the mountains to the east and prevents either vertical or lateral dispersion of pollutants. The natural haze of dust particles and ocean spray is augmented by smoke and other gases and, as has been shown by Meetham (1964), particularly gases which are subject to chemical changes as a result of long exposure to intense sunlight. Fine particles tend to accumulate just below the inversion layer and so conditions are worst in calm weather when the base of the inversion layer drops towards ground level. Similar problems are experienced elsewhere along the Pacific coast of the United States of America.

III Techniques of measurement and analysis

1 *Measurement techniques*

Frequently, existing data and information are not sufficient for adequate analysis particularly since, as Chandler points out, there is a considerable problem as to whether a recording site is typical for the pollutant in the area around the site. This latter problem in an urban environment has been examined by Weatherley (1963) and a method for dealing with it is presented. Another problem, pointed out by Chandler, is that the distribution of local authority recording sites is rarely even, and this is particularly so in sparsely-populated areas outside towns where few recordings are available. An uneven distribution of recording sites makes cartographic analysis difficult and can make statistical analysis impossible since many tests require a fairly regular coverage of data. The answer in many such instances may be to set up separate equipment in the sites that are required for the analysis, as was done by Garnett for the surveys of air pollution in Sheffield.

There are as many methods of measuring air pollutants as there are pollutants themselves, and so it is not possible to deal with many of these methods here. However, the problem in each case is basically the same, and that is to assess the concentration of the pollutant found in a given quantity of air. The exceptions to this are in the case of large solid particles which are deposited, and these are measured by weighing material deposited from the atmosphere. The methods of measurement vary considerably from relatively simple sampling processes to the use of complex chemical reactions.

Smoke is usually measured by collecting a sample of the suspended particles from a known volume of air and in some way measuring or assessing the weight (usually very small) of material collected. Sulphur dioxide is generally measured by passing a known volume of air through a known quantity of a specified liquid and assessing the chemical change resulting. A fairly simple method for measuring the quantity of suspended hydrocarbons is mentioned in the O.E.C.D. report (1964) and is effected by bubbling samples of polluted air through cyclohexane and comparing the fluoresence of the resulting solution with that of a standard solution of anthracone.

The most usual method for measuring smoke is frequently combined with a method for measuring sulphur dioxide. Air is sampled through an inverted funnel placed at least one metre from the wall of a building and at about first floor height. The air is drawn through a machined metal clamp of specific diameter between the segments of which is held a sheet of filter paper of a specific grade (usually Whatman No. 1). The particulate smoke is taken out of suspension in the air onto the surface of the filter paper, which is left in the clamp for a known period of time if sampling is continuous as is generally the case. The air is then passed from the clamp into a dreschel bottle containing hydrogen peroxide solution (at 1-vol. strength and pH 4·5), and is bubbled through the solution. The air then passes from the dreschel bottle through an air-flow meter (usually of the simple gas meter type) and to a suction pump which maintains the flow. The quantity of smoke collected is assessed by measuring the optical density of the particles collected on the filter paper. As both the volume of air flow for the period selected for sampling and the diameter of the clamp are known, the optical density (measured in terms of reflectance) can then be calibrated using a formula to give the standard concentration of smoke usually represented in microgrammes per cubic metre of air.

The concentration of sulphur dioxide is obtained by titrating the resulting solution in the dreschel bottle with sodium carbonate solution (at 0·0 N.) to determine the quantity necessary to neutralize the acidulated peroxide solution. This can then be used to calculate the quantity of sulphur dioxide. In this process it is important that the sampling tubes should be of a specific bore (usually 8 mm diameter if the flow rate is to be 2 m^3 per day) and that the funnel should be of a specific mouth diameter related to the flow rate (between 3 and 5 cm for a flow rate of 2 m^3 per day). In addition, the tubing should not be of either polythene or rubber since rubber absorbs sulphur dioxide and polythene becomes charged and so attracts particles in the air. The calibration of the results for smoke is related to the flow rate so that it is best to keep to the standard rate (2 m^3 per day). It will be clear from the above that frequently in this field measurement is best achieved by collaboration with specialists from other fields in order to avoid grave technical blunders.

2 *Analytical techniques*

(*a*) *Specific analyses.* As in most specialized branches of science, some methods of analysis are developed specifically for that subject, whilst others (e.g. statistical analysis) are common to most. The nature of some of the techniques developed to study air pollution is familiar to most geographers. In the examination by Clifton, Grimoldby and Sharp (1969) of areas in South-East Lancashire, an estimate of the domestic source density was made in the area immediately around each recording site. The method adopted was to place a circle of 1 km radius with its centre on each site on a 1:25,000 scale map. Within this circle an evaluation was made of the percentage of the area falling within each of four categories of source density. The percentages were weighted in the ratio 4:2:1:0 for the range from the highest source density to few or no pollution sources. This allowed areas without sources to be ignored and those with medium and high densities to be given more importance. The index of domestic sources could then be put on a scale from 0 to 10 by dividing by 25. Smoke control areas were catered for by rating them one category lower than they should have been for the housing density. As a result of such an analysis it was shown that the continuity of a particular source density over an area, or the lack of it, can be important in affecting the concentrations of smoke recorded. It was also found that local sources, such as roads, or poor dispersion near industrial sources, were a strong influence on the records from a number of sites.

In the case of sulphur dioxide pollution in South-East Lancashire, these same authors found that the highest concentrations were in the zone including the industrial and commercial areas of the major towns. A similar type of analysis for a sulphur dioxide source index was performed, using the same kilometre circles and rating the industrial sources within the circle by the size of installation (e.g. small—up to 15,000 lb/h steam, and large—over 100,000 lb/h steam). In fact only small and medium categories were used, since the large sources were not considered likely to have much effect on a site within one kilometre. The two categories were rated in the ratio 4:1. When there were multiple sources the circle was hatched on a diagram and these were evaluated by comparing with areas where the range of sources occurred separately. The index was again expressed in the range 0 to 10. This illustrates a cartographic method of analysis linked to social and industrial variables developed for a specific study of air pollution.

Where specific problems arise, then, particular methods of analysis often need to be derived to make proper assessment of the problem. This is illustrated by the analysis of Bach (1971) who examined the vertical and spatial patterns of air pollution in Cincinnati. Using the effects of particles in the atmosphere in scattering solar radiation, he calculated an atmospheric turbidity index based on the theories of aerosol physics.

(b) *Statistical analyses.* On the other hand there are a wide range of statistical techniques which, if they are used with caution, can be of considerable advantage in the analysis of the spatial aspects of air pollution research. One such technique is trend surface analysis. This is basically a surface fitting method allied to regression analysis, and an outline of the technique may be found in Chorley and Haggett (1965). Anderson (1970) has used this technique in the analysis of urban air pollution in the West Midlands conurbation. A surface of a particular power order, represented by a polynomial equation, is fitted to a set of spatial data so that two dimensions represent the position of each observation and a third represents the magnitude of the observation. The mathematical surface which may be fitted to the observed data may be of any polynomial order. It may simply be a linear surface, which is a plane, though not necessarily a horizontal surface, or it may be a higher order surface whose form is more complex. When trend surface analysis is used considerable interest is usually fixed on the analysis of the residuals, or deviations. These are the amounts by which each observation fails to sit exactly on the mathematical surface. It is assumed, not always correctly, that the surface depicts the general trend in the observations, while the residuals may be large enough to detect significant differences from the regional trend.

Anderson used surfaces based on cubic functions of the coordinates to construct generalized patterns of smoke concentrations for the West Midlands area. This, it was considered, allowed the examination of the general trend of smoke concentrations over the whole area, and maps are shown of the results. Fig. 8.4 shows first to third order surfaces of smoke concentrations on 6th January, 1965, expressed as percentages of the January means, in the West Midland conurbation. The results were categorized by days with wind predominantly in particular directions. The analysis showed that, as the wind speed increased on days with westerly winds, the local maxima of both smoke and sulphur dioxide tended to be dispersed and the pattern for both pollutants became much more even. There is some doubt, in this as in several other spatial studies, about the validity of using trend surface analysis in this role, owing to the fact

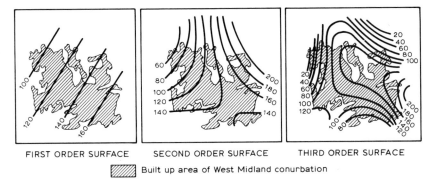

FIRST ORDER SURFACE   SECOND ORDER SURFACE   THIRD ORDER SURFACE

Built up area of West Midland conurbation

Fig. 8.4 First- to third-order trend surfaces for smoke concentration of January 6th 1965, as a percentage of the January means in the West Midlands conurbation. After Anderson (1970).

that it is difficult to establish a viable hypothesis as to which order surface should be fitted. If certain recognized constraints are followed, however, much useful information can result from the use of the technique. This in itself demonstrates the care which must be taken to select a suitable statistical technique if one is to be used, but having selected an appropriate technique many useful results may be obtained.

The geographer, then, should be in a position to assess fully the potential effects of the landscape, in the broadest sense of the term, on the distributions of air pollutants in the atmosphere within a particular area. This assessment can aid him in setting up hypotheses for further analysis for which he has a variety of tools at his disposal, both of a strictly statistical nature and of a more specific nature, designed to fit the particular analysis.

3   *Analyses of the diffusion of pollutants*

Having examined the relationship between the distribution of air pollution and the physical and social environment, the next step would appear to be an examination of the diffusion of pollutants from a source area. Such a study is fraught with problems, since the diffusion of pollutants is closely linked with the meteorological conditions prevailing at a particular instant. Some examination of the diffusion of pollutants is provided in Scorer (1968—especially Ch. 2), and Meetham (1964—especially Ch. 12). From the theoretical point of view, it is simpler to consider the diffusion of particles or gases from a single source such as a chimney plume rather than from a whole area (e.g. an area of factories or dense housing). Scorer provides a basic mathematical treatment of the single plume case.

In the analysis of urban air pollution in the West Midlands (Anderson, 1970) already mentioned, the use of trend surface analysis was also directed towards an examination of the diffusion of pollutants. The residuals from the third order surface were examined on the basis that the consistency in the magnitude of residuals at a particular site for days with a given meteorological situation indicate particular parameters operating in respect to diffusion for the area. If such a view were acceptable, the results could be applied as a correction to the concentration of pollutant predicted by a general diffusion model for the site. It was found, however, that insufficient consistency of magnitude and sign were obtained for statistical inference for particular weather conditions. But when triangular areas were set up embracing particular kinds of land use upwind of a recording station, and with the apex of the triangle at the recording stations (using triangles with vertical angle 45° and height 800 m), the relative residual (that is, the residual computed as a percentage of the base value) was broadly consistent for particular types of land use within the triangle. For example, it was found that the stations with positive residuals for smoke were situated in areas containing large amounts of residential development, while stations with consistently positive residuals for sulphur dioxide showed a mixture of housing densities and land uses.

Martin and Barber (1967) examined concentrations of sulphur dioxide at various distances from High Marnham Power Station, Nottinghamshire (see Fig. 8.5) during the period October 1965 to September 1966. The study considered pollution from background sources in the area as well as from the power station itself. A survey was made in the area around West Burton power station some 15 km north of High Marnham, prior to the commissioning of the latter station, to study existing pollution in the area. There were no significant emissions of sulphur dioxide from West Burton. The survey sites form 3 rings of 8 stations at 5, 9 and 13 km from West Burton; and a tower for meteorological instruments was also present standing 46 m high, so that atmospheric temperature gradients could be measured, between 6 and 46 metres. Martin and Barber plotted the variation of average sulphur dioxide concentration with wind direction by the use of pollution rosettes for each site, both for winter 1965–66 (see Fig. 8.5) and summer 1966, and showed that most pollution comes into the area from the north-west, west and south-west with the possible sources being Leeds, Sheffield and Nottingham.

There were problems in measuring the concentrations from a

single source such as High Marnham power station since these varied considerably with:

(i) the period for which the readings were averaged
(ii) the steadiness of the wind during this period
(iii) the variation of the wind with height.

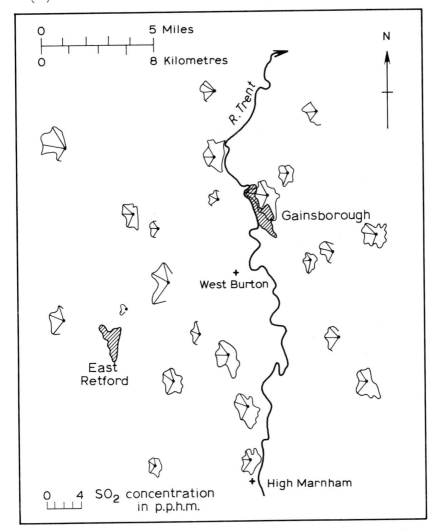

Fig. 8.5 Pollution rosettes showing the variation of average sulphur dioxide concentration with wind direction at each recorder site around West Burton power station (winter 1965–6). The amount of pollution received is shown drawn from the centre back along the direction from which the pollution arrived (pphm—parts per hundred million). After Martin and Barber (1967).

It was concluded, however, that less than ten per cent of the pollution recorded during the period October 1965 to September 1966 came from the direction of High Marnham. In general, Martin and Barber found that high concentrations of sulphur dioxide from High Marnham could only be detected on a short-term basis. Sulphur dioxide was detected up to 27 km from this power station, but only very infrequently, and no concentrations from this source were detected to that distance under stable atmospheric conditions. Martin and Barber concluded from their analysis that pollution from High Marnham arrives at sites in a wave form with amplitude and frequency decreasing with distance, assuming a steady wind.

These, then, represent two studies of the diffusion of pollutants. The first was based on an analysis over a fairly wide area and involved considerable generalizations, while the latter was a relatively intensive study of a small-scale phenomenon which nevertheless results in implications about the pollution arriving from other sources. Study of diffusion of pollutants brings the geographer into the study of process—the process by which pollutants are disseminated whether over short or long distances. Increasing knowledge makes possible better and better models of what is going on and, as Anderson (1970) implies, the geographer may hold the key to the application of these models in particular locations and settings.

IV The control of air pollution

1 *Legislative controls*

(*a*) *General considerations.* Much has been done in recent years in Britain, partly under the influences of the 1956 Clean Air Act and preceding legislation, to control the output of pollutants into the atmosphere by the improvement of industrial plant and of domestic heating applicances. Much has also been done in other countries. In America control legislation is complex since it is in the hands of individual States (Magill, Holden and Ackley, 1956). These developments, as such, do not fall directly into the field of interest of the geographer, but the geographer might well be of value in suggesting where the legal provisions might be implemented.

Holford (1958) examined the importance of clean air in urban planning in a lecture entitled 'Clean air, urban amenity and the growth of towns'. He reviewed the impact of the desire for clean air in relation to the range of spheres of planning activity: countryside, housing, public services and utilities, industry and commerce, and social provisions. In the first he emphasized the danger of the drift of pollution from urban areas, regarding the country as an area of urban amenity, and the deleterious effect of pollution on forestry and farming. In the case of housing, he noted that at the time planners were still siting new housing estates in frost pockets and in areas subject to frequent temperature inversions, and suggested that planners should pay particular attention to the characteristics of new land for housing in relation to clean air. In the case of public services and utilities he emphasized the effect, at the time, of most forms of transport on the incidence of fog and suggested that abatement of traffic pollutants should be encouraged as much as possible. He reviewed town planning practice in relation to the location of industrial and commercial functions, and concluded that more selection was needed in the establishment of groupings of industries in particular areas. In relation to social provisions he mentioned the dilapidation of public buildings, such as cathedrals and monuments, as a result of air pollution, and how much money might be saved if restoration was unnecessary.

(b) *Policies in Oxford and Cambridge.* There is a wide range of planning activities, however, in which the problems may be numerous, and Holford illustrated this with the methods adopted by Oxford and Cambridge. Both cities are free of drift pollution from industrial districts in the vicinity, and both are subject to high humidities. Oxford, however, has a generally concave land shape and so suffers from downward-flowing breezes which, during anticyclonic weather, give temperature inversions, and a tendency to low level fog. Both cities had a considerable air pollution problem; their planning policies in this respect, however, diverge. Cambridge deliberately limited its urban spread and encouraged residential and industrial expansion in surrounding villages, tending to reduce the pollution problem in Cambridge. Oxford seemed to accept the expansion as inevitable and so would have to build its new housing and industrial estates as smoke-control zones.

The above suggests that there is considerable scope for geographers to work with planners. The geographer should be valuable in assessing the factors controlling the distribution and dispersion of pollutants and thus in suggesting areas for development of different types of land usage, or in the assessment of the most advantageous areas for the establishment of smoke-control zones.

2 Smoke-control zones

The most widely effective legislation in Britain with regard to air pollution is concerned with smoke emission, although many toxic chemicals which might otherwise be emitted from industrial plant are prohibited. The 1956 Clean Air Act in Britain is outlined and discussed by Garner and Crow (1964). The principal concern of this Act for geographical study is the provision for smoke-control zones. This empowered local authorities to establish zones in which the emission of smoke from any chimney is prohibited except if certain administrative requirements have not been carried out, such as the failure of the authority to warn a particular occupant of the impending restrictions in an area, or if certain prescribed fuels did not function as intended. Crown premises are exempt from the provisions of the Act as a mandatory stipulation but are encouraged to eliminate the emission of smoke as far as possible.

Fig. 8.6 shows smoke-control zones in the Nottingham, U.K. area up to 1969 and shows that the Central Business District, where there is little residential development, is a smoke-control zone. Around the north and west perimeters of the city, and in the area to the south and west across the river Trent, are areas of fairly recently built estates except for small patches of old centres such as Bulwell. In these areas, as can be seen, smoke-control zones have been set up; in such

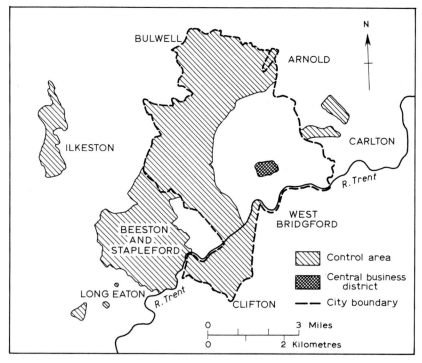

Fig. 8.6 Smoke control areas in Nottingham in April 1969.

cases it is simple to impose restrictions since the housing is new and more easily modified to new heating appliances. This area is also perhaps significant since it protects the city from pollutants which might be brought in by westerly winds. Outside the city the policy varies considerably from one local authority to another, and from one area to another. Nearly the whole of Beeston and Stapleford is covered by control areas, and shortly the whole will be, while in Long Eaton only relatively small areas have been designated.

The selection of zones suitable for smoke control involves the consideration of a large number of factors, but it would seem obvious that one of them ought to be an evaluation of whether the establishment of control zones in a particular area will be advantageous in controlling pollution in the environment as a whole. One might well investigate the establishment of control zones in low-lying housing areas frequently subject to local inversions, while not being so concerned about well-ventilated areas. The establishment of control zones to leeward of parks and open spaces to extend the advantages created by the filtering effects of vegetation might well be considered. In this sort of respect the geographer would be usefully employed working with others to improve the environment.

It is demonstrable that the geographer has a role to play in the analysis of the problems faced as a result of air pollution, in subsequent planning to alleviate the problems and, more importantly, in catering for the future so as not to let lack of foresight in present planning create further problems. The problem orientation of the geographer's work should need no further exemplification in this field. In a highly technical subject, such as air pollution, a geographer can certainly not be expected to amass sufficient expertise to cope with the whole subject on his own; only by working on particular specialized problems and in close collaboration with specialists from other disciplines can he produce useful and applicable results.

References

ANDERSON, P. (1970): The uses and limitations of trend surface analysis in studies of urban air pollution. *Atmospheric Environment* **4,** 129–47.
BACH, W. (1971): Atmospheric turbidity and air pollution in greater Cincinnati. *Geographical Review* **61,** 573–94.
CHANDLER, T. J. (1965): *The Climate of London.* London.
CHORLEY, R. J. and P. HAGGETT (1965): Trend surface mapping in geographical research. *Transactions of the Institute of British Geographers* **87,** 47–67.

CLIFTON, M., M. R. GRIMOLDBY and A. J. SHARP (1969): *Air pollution in South East Lancashire: A preliminary study of the winter pollution data from the National Survey 1961–62 to 1967–68.* Warren Spring Laboratory.

CRAXFORD, S. R. and M. L. P. M. WEATHERLEY (1968): *Air pollution in Great Britain.* Warren Spring Laboratory.

(1971): Dispersal of airborne effluents: air pollution in towns in the United Kingdom. *Philosophical Transactions of the Royal Society, London series A,* **269,** 503–13.

GARNER, J. F. and R. K. CROW (1964): *Clean air: law and practice.* London.

GARNETT, A. (1967): Some climatological problems in urban geography with reference to air pollution, *Transactions of the Institute of British Geographers* **42,** 21–43.

(1970): Sheffield emerges from smoke and grime, *Geographical Magazine,* **43,** 123–8.

HOLFORD, W. (1958): Clean air, urban amenity and the growth of towns. *National Society for Clean Air, Proceedings of the Llandudno Conference 1958.*

MAGILL, P L., F. R. HOLDEN and C. ACKLEY, editors (1956): *Air pollution handbook.* New York.

MARTIN, A. and F. R. BARBER (1967): Sulphur dioxide concentrations measured at various distances from a modern power station, *Atmospheric Environment* **4,** 655–77.

MEETHAM, A. R. (1964): *Atmospheric pollution: its origin and prevention* (3rd edition). London.

ORGANIZATION FOR ECONOMIC COOPERATION AND DEVELOPMENT (1964): *Methods of measuring air pollution.* Paris.

SCORER, R. (1968): *Air pollution.* London.

TURNER, W. C. (1955): Atmospheric pollution. *Weather* **10,** 110–15.

WEATHERLEY, M. L. P. M. (1963): Interpretation of data from air pollution surveys in towns taking into account the siting of the instruments. *International Journal of Air and Water Pollution* **7,** 981–7.

9 Health

Neil D. McGlashan

I The relationship of geography with medicine

The rapid growth of the human population is accepted today as an
inevitable and gloomy fact. Few people appreciate that it has at least
one beneficial, although by no means fully compensating, advantage
for mankind. The range of man's collective understanding should
increase in step with his absolute numbers. This too has its consequence. Academic specialization is inherently bound to increase.
The specialist will indeed know more and more about less and less.
Many of these specialists will be likely to undertake research in the
interstices, so to speak, between accepted disciplines. Medicine and
geography are two such classical disciplines whose area of common
interest has only attracted attention to any great extent in the last
two decades.

Medical geography is concerned with a study of those local variations of environmental conditions which are causatively related to
human health or ill-health. Hippocrates, writing in the fourth century
B.C., had a clear appreciation of this relationship and his essay
On Airs, Waters and Places places emphasis upon both physical and
social geography.

> He who wishes to study the art of healing must first and foremost
> observe the seasons and the influence each and every one of them
> exercises . . . and further he shall take note of the warm and cold
> winds . . . so should he also consider the properties of the water . . .

The healer shall thoroughly take the situation into consideration and also the soil, whether it is without trees and lacks water, or is well wooded and abundant with water, whether the place lies in a suffocatingly hot valley or is high and cool. Also the way of life which most pleases the inhabitants; whether they are given to wine, good living and effeminacy, or are lovers of bodily exercises, industrious, have good appetites and are sober.

Climate, locality and human activities are thus all looked upon as potentially causative of ill-health. The human factors arise both from economic occupations—say, chimney sweeping or rubber manufacture—and from popular recreations—say, mountaineering or usage of tobacco. The interrelationship of these factors is also implied. Asbestos miners' lung cancer depends upon both local geology and human exploitation of that resource.

Although a long ancestry—and, by implication, reputable standing—is thus indicated for medical geography, little advance has been made by the subject over the centuries. The reason lies very largely in technology within the two parent disciplines. Geography has only recently emerged from the morass of descriptive compendia of new discoveries—'far-away places with strange-sounding names'. Medicine has progressed only quite recently from astrology and the four humours to the discovery of the circulation of the blood and minute studies of bacteria, virus and cell structure. Neither discipline was equipped to undertake broadly-based studies depending upon quantitative assessment of geographical factors and their impact upon human population groups. For instance, observation 'of the seasons and their influence' depends upon the invention of the thermometer; 'the properties of the water' can be geographically studied only when chemical trace-element techniques have been developed (Warren, 1964). Spatial variations of disease occurrence cannot be analysed until basic demographic information is available and until scientific standards of medical diagnosis and recording are set. Hippocrates' ideal sources of information also depend upon contributory data from many other physical and social sciences: meteorology, pedology, botany, for example, or sociology, anthropology, economics.

Even when each science has established its own standards of technical measurement, data are often not everywhere available on each one. Much of the world, for instance, still lacks soil survey information. At a later stage, by contrast, the sheer mass of scientific data becoming available threatened to swamp the geographer's wish to consider *all* possible factors in his analysis. He had no means of handling the volume of information—a true *embarras de richesse*.

Two modern developments are today tending to combat this analytic inadequacy. First, a similar interdisciplinary bond between statistics and geography has given rise to development of quantitative methods of multi-factorial appraisal in medical, as in other branches of geography. The basic concept here hinges upon the recognition of groups of factors significantly co-extensive with areal patterns of health or ill-health parameters. The methodological framework for such studies within geography stems from regional analysis where relationships are sought, in geographical space, between diverse factors. The culminating point is to recognize causation; a plus a little b in the presence (or absence) of c causes d. Applied to medical problems this is aetiology. Its importance lies in identifying previously unrecognized causative factors as a preliminary to removing or controlling factors harmful to man. The handling of these diverse types of quantitative data is a problem general to all geography's many applications. What matters is that scientific repetition of results leads on to confident inference of general truths.

Secondly, the invention of the electronic computer, with its constantly increasing power and storage capacities, provides for the geographer a tool to carry out laborious calculations (Armstrong, 1972a). He has become free to conceptualize fields of analysis previously quite unfeasible because of the total of lifetimes needed for calculation by human means. The costliness of such equipment may be argued; but even the humble desk calculator makes possible great improvements on the old descriptive geography. Thus, and only in the last twenty years or so, have quantitative methods and the computer era combined to make feasible work on the lines formulated by Hippocrates' school of physicians.

II The data of medical geography

1 The character of the data

Health has been defined as a state of full physical and spiritual well-being and of perfect adaptation to environmental conditions (May, 1958; 1961). Yet it is a paradox of medical geography that studies are almost invariably of ill-health, disease and death. This is perhaps, to study health negatively by looking at its converse side. Many conditions of ill-health may be looked upon as a continuum from overt sickness through slight malaise to normal healthiness with peak conditions of physical fitness attained only briefly in most human lives. The point at which disease is identified may thus be fairly arbitrary. Nonetheless, data available for study is upon ill-health rather than positive health.

It is convenient to consider the sources from a geographer's viewpoint in four groups and, in passing, to stress that specialities within geography stem chiefly from the type of data most often utilized. Daily working with a particular group of space variables, whether industrial, geomorphological or medical, leads to greater understanding and facility and hence to specialism in their use.

Basic medical diagnostic information upon ill-health or cause of death is rarely collected in a strictly geographical sense. Place of exposure to environmental stimuli or stresses is often inadequately recorded. A primary source of information is the death certificate, but this is not without its problems. A cancer patient may die in a car crash. A slow or non-fatal illness may be swiftly overtaken by death from an acute but unrelated cause. It may be desirable to look beyond the primary cause of death to other potentially serious ailments within the same cadaver (Case and Davies, 1964). These often remain undiagnosed and unrecorded. Only an atypical group of corpses, possibly of forensic or special medical interest, are sent for *post mortem* examination. Other factors too influence the cause of death recorded. The area of special interest or skill of the physician, his consideration for surviving relatives, even semantic fashion, may influence the recording of the cause of death.

Morbidity data have their own difficulties. Confidential patient/ doctor relationships may be breached by allowing research access to personal medical details. Some conditions may be socially inadmissible, extramarital pregnancy or psychiatric disturbance for instance. Diagnoses may alter after first recording; patients may suffer from several conditions simultaneously, especially in the underdeveloped world, and be recorded only under the most treatable one. Diagnostic criteria may vary personally or spatially and treatment methods may vary in success rates. From the patient's point of view his perception of ill-health and need to seek medical advice certainly varies with distance, cost, and even religious flavour imparted with the treatment. Allowances can be calculated for many of these data difficulties but the medical geographer must be alert to avoid reaching quite erroneous conclusions based upon uncritical use of specialized subject data.

The second main group of data sources is statistical. Census information on population with age, sex and especially place breakdown is essential for comparative spatial analysis. Statistics on birth and death rates, fertility and infant mortality, are still inadequate in much of the world. Only with this type of information can standardizing procedures be applied to allow valid geographical comparisons. Often the administrative borders used for official statistics are out of date as cities spill into surrounding countryside. For example, an under-bounded urban area cannot be validly compared with one whose boundaries still include much open farmland. In under-developed countries some means must be found of utilizing less than ideal data in order that research opportunities present today are not lost by deferring enquiry until a more perfect data base exists.

Thirdly there is the geographical matter of locational accuracy of information. Hospital statistics, for instance, are usually based on an indefinable service area and consequently an unknown population size. Patients' real points of origin may be concealed by mere over-night addresses near to the hospital, or indeed no address at all may be recorded. Homes and workplaces may be in different hospitals' zones or in differing administrative or statistical collection divisions. Occupational risks may have been incurred in previous or in present employment or even, with increasing recreational travel, in an entirely foreign holiday environment. Workmates and social contacts form an immense potential diffusion net for acquiring and passing on sickness. All may need locating, even basic topographical maps will not be available in every country. Air-photograph coverage may sometimes be of value, and maps based on population weightings often assist towards a realistic representation. For instance, the dense population of midland Scotland compared with the Highlands is better represented in Howe's second mortality atlas (1970) than in his earlier one (1963). Forster (1966) has used a locational cartogram for each sex and age group to give a similar, demographically more correct, impression than does a normal topographic map.

Fourth is the general problem of choice confronting the geographer, who must ideally consider or utilize *all* available information. In practice of course he selects and, moreover, selects on an intensely subjective basis from an often vast store of potentially mappable information. Government agencies and technical or charitable organizations frequently collect and store a great variety of information. Much of this will need considerable processing before being suited to medical geographical analysis; much will be irrelevant and much will be totally lacking. Variations of agricultural production may be obtainable; details of daily diet may not be. Often collections for a different purpose may be unsuitable for geographical analysis. Data from many commercial undertakings are of this kind. The author recalls having had to try to follow through the distribution chains of chemical agricultural fertilizer in hypothesized connection with stomach cancer, and of bottled beer in connection with spatial variation of road accidents.

Alternatively the medical geographer may be compelled to collect the information he requires—particularly he will often want information on local customs, likes and dislikes, even opinions and perceptions. Thus, as most geographers do, he has to devise a specific enquiry, perhaps by questionnaires, for his purpose. He may have to do his own field surveying, for instance, in West Africa to find the various horizontal and vertical distances of village sites from water for comparison with data on onchocerciasis (river blindness) occurrence by hut and flight distances of the vector, *simulium spp.*, flies. Acland (1856), studying cholera in Oxford in 1854, was fortunate in obtaining from an engineer a map of the city contoured at five feet vertical interval.

This fourth section of miscellaneous, possibly complementary, data makes up what May (1954) has called geogens—'factors known or thought to have a (positive or inverse) relationship with human health'. Note the subjectivity of the definition. What one man *thinks* to be relevant drawing on his personal experience may, to another, seem merest whim or fancy. Nonetheless, from thoughtful speculation truths emerge.

2   *Data presentation*

The difference between a statistical map and a table of the same statistics lies in the map's ability to show the relationships of areas. Like values may cluster in space or be geographically dispersed. Petermann (1852), writing during what has been claimed to be the golden age of medical cartography (Gilbert 1958), explains the point well.

> The object, therefore, in constructing Cholera Maps is to obtain a view of the Geographical extent of the ravages of this disease, and to discover the local conditions that might influence its progress and its degree of fatality. For such a purpose, Geographical delineation is of the utmost value, and even indispensable; for while the symbols of the masses of statistical data in figures, however clearly they might be arranged in Systematic Tables, present but a uniform appearance, the same data, embodied in a Map, will convey at once, the relative bearing and proportion of the single data together with their position, extent, and distance, and thus, a Map will make visible to the eye the development and nature of any phenomenon in regard to its geographical distribution.

The map remains one of the main tools of geographical interpretation today—the stock-in-trade of geographers. However, the maps compiled by medical geographers are often for the instruction or action of physicians untrained in any cartographic awareness. It is therefore more than ever important that the quality of data utilized and the limitations of mapping conventions be absolutely explicit. To non-geographers a printed map assumes canonical authority! We must ourselves beware of conveying spurious reliability.

Today's hectic speed of life and the advent of the computer line printer has added another aspect to medical mapping. The map may be merely a print-out of symbols or signs correctly located on an assumed but undepicted topographic base, with none of the fine lettering, whales, ships and elephants of previous ages. Such maps, made in and for one moment, serve for analysis and are gone. These would only be actually hand-drawn if needed for permanent display (Hopps, 1968). At the other extreme of sophistication the computer can produce fully calculated isopleth maps of an individual disease condition based on age and sex standardization—a lengthy and tedious operation by hand. Computer mapping is also well adjusted to use with a grid. Armstrong (1972a) has used it to plot reported health hazards (for example rat infestation, derelict cars, etc.) in an urban complex and showed a dramatic spatial similarity with infant deaths. Similarly traverse data is easily computerized and so is the *pseudo-traverse;* the map of river or road pulled out, like a piece of crooked string, into a straight-line traverse for computer-based analysis.

Increasingly today geographers are using other forms of display in addition to the map. Many statistical diagrams and histograms and, especially, graphs are chosen for their convenience in showing several factors varying through time—for example infant mortality, population density and air pollution, each on its own vertical scale, all against time on the horizontal axis. Graph presentation is increasingly used for regression lines with significantly varying limits, high and low on each side (see also Ch. 4, section I. 2).

Whatever the medical geographer's selected mode of presentation, others will use the data so portrayed, which must therefore be in-capable of being misunderstood. Learmonth (1972a; 1972b) has recently most usefully reviewed the ways in which medical men have presented and analysed geographical data, and he has also described some of the many ambitious attempts at disease-atlas compilation by geographers, often working in teams.

III Spatial analysis in medical geography

1 *The location of medical facilities*

Geographers are constantly preoccupied with decisions upon opti-mizing location. In no field is this more appropriate than in medicine, where the cost of facilities today runs into millions, whatever the unit of money. Moreover, though this is not ideal, British patients are still using hospitals positioned by Miss Nightingale. The longevity of hospital structures is likely only to increase. There is clearly an absolute imperative to site these costly, quasi-permanent items of public expenditure best for that public. The argument applies whether the item is a school, bank, co-operative tractor station, maternity or dental clinic.

The simplest case, because sparse data at that time ruled out a more complex approach, may be illustrated from Central Africa (McGlashan, 1968). All hospitals in Malawi were mapped with their sphere of influence, defined in terms of actual patient travel. Each hospital's total medical facilities were assumed to be proportional to its in-patient bed numbers for general use. The hospitals' spheres of influence were superimposed on the dot map of the latest census of population and a simple calculation made of hospital work load in terms of ratio of facilities to population potential within the area. In contrast to expectation it was found that the rugged northern region was normally well served, but that three neighbouring hospitals in the centre of Malawi were potentially overloaded. The most ill-served population was in an area notorious for opposition to the President. Prognosis poor . . . .

A similar study (Jackman, 1972) was undertaken by geographers for the Zambia Flying Doctor Service, whose problem was to bring its clinics to within 56 km radius of every native not already within that distance of a resident doctor. In this case, undertaken after only one full census, current population distribution was considered together with roads to service the airstrips, existing rural clinic buildings and lake-based floating clinics. The result was an ordered recommendation for building airstrip clinics based on priority for greatest potential need.

In similar vein, Cook and Walker (1967) have studied regional variations in dental care in the United Kingdom at two dates ten years apart. When compared, the results showed a marked and unimproving regional imbalance of both dentists and their rewards. Possible measures were discussed with the object of alleviating a situation inequitable for both dentist and public.

The classic in this field, now widely emulated, was Godlund's study in Sweden in 1961. His purpose was to identify certain district hospitals which should receive the vast capital expenditure needed to upgrade them to become centres of regional medical services. Godlund considered population growth and decay trends and public transport facilities for a 15-year forward period, and developed a measure of people and travel times about potential regional centres to select those which would become best placed.

From Chicago recently have come a series of studies taking God-lund's fairly crude but basic assumptions rather further (Morrill, 1966; 1967; Earickson, 1970; de Vise *et al.*, 1969; Pyle and Rees, 1972). The essence of some of these studies has been to match each particular type of patient with a vacancy in the type of medical facility he needs (Morrill and Earickson, 1968; 1969). Studies have also been made of prejudices of patients by which they may opt to travel to less convenient facilities of their own religious persuasion. Ethnic divisions in the nation make an eightfold difference in expected travel distances to hospitals for negro compared with white patients (de Vise *et al.*, 1969). Pyle (1971) has recently gone a step further by selecting certain major killing diseases—heart disease, cancer and stroke—and extrapolating their rates at current trends into the future to match with expected facilities. His techniques are complex and time-consuming. However, this sort of practical interest in man's future health and well-being is a clear role for medical geographers.

The practical role of geographers in the administrative and spatial problems of medicine are not tied solely to the overdeveloped city. Closer to nature, Prothero (1965; 1968; 1969) has, in several terms as a W.H.O. consultant, studied the disease problems of Sudanic Africa, especially those related to human mobility in this harsh environment. Again, the prognosis for independent Africa is not good: although clearly interdependent in health matters, the nations are too often prone to political division.

Finally, in this group of medical geographical studies of administrative value, mention must be made of a recent study from Newfoundland, for it is basic to all attempts to provide treatment for the sick. Girt (1971) has reported on the ways in which distance from treatment affects patients' perception of ill-health in themselves. He finds, in a continuing study, that it is not those nearby nor those far from help who most readily react to their own symptoms by seeking treatment. Although behaviour varies somewhat for different diseases, it is most generally the middle-distance patients who perceive ill-health soonest. In Newfoundland this is a distance of about 19–24 km from a cottage hospital but, as Girt points out, the distance scale will be expected to vary in other environments. Since distance to treatment alters our demand for treatment, new facilities based on present or forecast demand may be proved seriously incorrect (King, 1962).

Disease surveillance for public health purposes involves defining incidence in different population groups in such a way as to emphasize significant changes in demographic and geographic characteristics of disease as these occur through time. An effective way of doing this is to map a series of time periods for a certain condition and follow, so to speak, the wax and wane of its spatial patterns. This theme will be considered again in the next section on spatial definition of disease but, for the U.S., Armstrong (1972b) has reviewed the gamut of skills required for this intensely geographical task.

2   *Spatial definition of ill-health*

The methods of defining areal variations of disease or death occurrence on maps are much the same as for other geographic phenomena. In general more sophisticated mapping techniques are appropriate only for higher quality data; but the common assumptions of geographers about their maps may need questioning since these medical maps will be utilized by persons who do not necessarily accept or follow our cartographic conventions.

Probably the simplest distribution maps involve the use of dots. John Snow (1849) used this form at a large scale to show every individual cholera death in a part of London in 1848. Clusters of deaths were shown up in close proximity to a particular public water supply pump. This map form has been used by physicians to show oesophageal cancer in the Transkei (Rose, 1965) and to show blindness in (then) Northern Rhodesia (McGlashan, 1966). Dots are appropriate where absolute case numbers are to be stressed and there is a location for each occurrence. Even so, a dot map has the danger of showing merely a pale shadow of the total population. The dot map is less successful, because less obviously interpretable, when scale requires amalgamation of dots representing, say, 50 cases, especially if amalgamation is within set administrative boundaries. Rarely then can the super-dot have much locational meaning (Rose, 1965).

The next step forward is to relate case numbers (disease or death) to the total population potentially at risk within an area. If required and if information permits, subdivision of the total population by age and sex will permit inferences to be drawn as between different cohorts of population. Changes in sickness rates over a period are related to surveillance, as previously mentioned. Fonaroff (1968) has neatly displayed declining rates of malaria in Trinidad by mapping spleen-rates over a period of years, and by examining historical records which bear out his observations over a longer period. Choropleths of death may also be based upon standardization procedures to allow for local variations of population structure. Howe (1963;

1970) has used this method in his atlases of mortality in the United Kingdom. This technique depends upon detailed census and cause-of-death data by age and locality and, of course, only certain countries can provide such detail.

Where less detail is available as, for example, in East and Central Africa, more approximate methods have to be devised. Cook uses various sites of cancer as a ratio of all cancers in the hospitals of East Africa (Cook, Collis and Foreman, 1971) and a geometric scale of category size for worldwide distribution of certain cancers (Cook, 1967). Palmer (1963) suggested approximate categories for disease definition in 1963, and this scheme was further developed for use in Central Africa in 1966 and 1967 (McGlashan, 1967). The concept hinges upon the time periods of the calendar and doctors' memories of disease experience. It expects a practitioner to be able to recall whether he sees a certain condition only once a year, once a month or more than once a week. The justification for using an admittedly low-quality data base is that useful results have emerged (McGlashan, 1969).

By contrast, in countries with detailed medical and demographic data Learmonth and Nichols (1965) showed that choropleth maps, treated as representing a continuous surface, may be developed into isopleths of disease or death—isomorbs or isomorts. For Australia, Learmonth and Grau (1969) published isomorts of selected causes of death with the caution to the user that these are contrived carto-graphic devices. His introduction emphasizes the conventions he uses and it is an explanation not to be lightly ignored!

By choosing from methods such as these, the geographer can pro-duce maps showing the spatial patterns of medical conditions. This is not enough. Any locational patterning will have inherent variations as a result of the operation of random chance. No explanatory analysis can be soundly based until the distribution studied has been shown to diverge at a significant level from that attributable to chance factors.

Choynowski (1959) perhaps pioneered the application of prob-ability testing to medical maps with his terse description of brain tumour mapping in part of southern Poland. Although fairly wide variations of rate of occurrence of this rare condition were mapped, only two administrative areas reached the 95 per cent level of signi-ficance by Poisson test. Since this test is suited to rare events it is most commonly used in this type of work where a death from p in area q in year r may be considered statistically rare. The same test has been applied, over a run of years, to male stomach cancer deaths amongst Europeans in South Africa (McGlashan, 1965) and, for various time periods comparatively, to leukaemia deaths in England and Wales

(White, 1972). Aggregation of data by time, of course, assists one to reach a total of occurrences large enough to produce significant spatial variations.

Fig. 9.1 depicts 47 municipal divisions of Tasmania (population 400,000) and male deaths from heart disease—a major killer—for a decade. Taking two five-year periods, 1958–63 and 1964–8, and the total decade severally, areas with significance (p > 95 per cent) have been counted for both high and low rate areas. (Areas with significant values in one of these periods are marked 1 in the figure, areas with significant values in two periods are marked 2, etc.) No area features as changing from high to low or low to high rated significance, and the overall effect divides the island into two clear halves with considerable indication of temporal consistency. Spatial and temporal consistency was also amongst White's major conclusions from his leukaemia work (1972). There is surely a hint here of some unchanging causative factor(s), and a clear lead to an environmental search for culpable phenomena.

Although tests based on the Poisson distribution have been the commonest ones to be applied in medical geography, chi-squared, standard deviation or other spatial tests of non-random occurrence may also be used, especially if larger numbers of cases are available. Besides temporal aggregation mentioned above, totalling of cases in neighbouring areas may in theory be used. This has the disadvantage of losing some of the geographical specificity of the original data, as indeed time totalling may conceal the effects of changes of disease rates or reflect changes of treatment or prophylaxis through time.

Armstrong (1969) has advocated the regular use of the standard deviation as the base for choropleth mapping of health situations. It would, he claims, provide uniformity and hence comparability of mapping. This has the advantage of combining the rate information of the choropleth map with the significance attached to plus or minus two or three standard deviations. Implications must be treated with extreme care, however, when numbers involved are small, especially since the standardized mortality ratio conceals the numerical facts from which it is calculated.

3   Associative occurrence in space

The most usual method for comparisons of distributions is the assessment of apparent similarity by eye. We are all familiar with similarities on continental scale of, for example, rainfall and vegetation maps. Statistical evaluation allows one to go further and to put the comparison upon an objective footing and to assess just how alike distri-

butions are and at what level such a similarity might not have
occurred by chance. Such comparison may be in pairs severally—a
disease with a socio-economic factor—or it may, with computer
facilities, run into multiple regression.

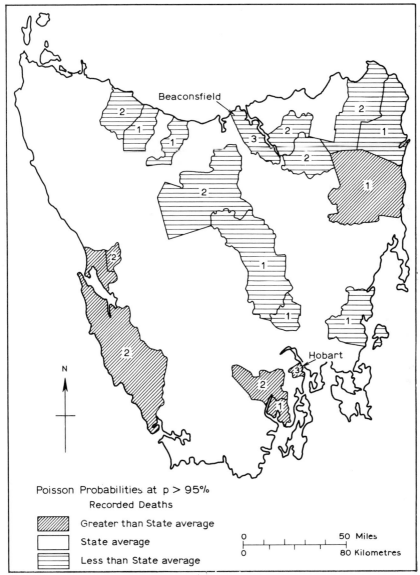

Fig. 9.1 Male deaths from Ischaemic Heart Disease in Tasmania 1958–
68—accumulated significance. Numbers refer to significant values in
one, two or three of the time periods 1958–63, 1964–8, 1958–68.

Girt (1972), working upon chronic bronchitis in Leeds, and Dever (1972), working with leukaemia data in New York, have compared disease distributions with a number of housing and other social variables. Girt's results emphasize the role of past risk factors in the environment, and Dever shows the varying results obtainable as the scale of focus upon a problem is varied.

This leads to the vital problem of recognizing causation. In a paper which compares merely pairs of variables, diabetes and cassava-eating, cancer of the cervix uteri and male circumcision (McGlashan, 1967), attention has been drawn to the care needed in interpreting geographical results in medicine. Since geographers study population generalizations, they will rarely prove a causative relationship, only perhaps draw attention to a previously unrecognized interrelatedness between variables. This method of approach has little to offer in the study of conditions whose aetiology is understood, but it may contribute much to those whose cause or causes still await elucidation.

Other interesting examples of this type of approach, not included here for reasons of space, are provided by Sakamoto and Katayama (1967) with studies of seasonal variations of disease incidence amongst racial groups in the United States. Another environmental factor which has been studied by Warren (1959) is soil trace-elements in relation to goitre and multiple sclerosis. As Armstrong (1964) has pointed out (in another context) the causative pathway from soil to man is a long and complex one, liable to misunderstanding at many points, especially in a world where few of us today eat predominantly, or even significantly, the products of local gardens.

4   Disease diffusion models

Another variable of vital concern to the geographer is time—the changing distributions of phenomena through time. A very simple example is the finding that some seasonal peaks of disease incidence become less marked through time (Sakamoto and Katayama, 1967). Another aspect is the cumulative effect of some disease-causing mechanisms through time—a case in point is the smoking/lung cancer relationship.

Rather more sophisticated in approach are studies of disease spread through time *and* space. For the geographer epidemic waves are similar as a topic to other studies of innovation diffusion in which, to borrow sociological language, one can identify innovators, adopters and immunes. An outward spread from a centre can be postulated, like ripples from a stone thrown into a pond. The advance of an epidemic wave can be mathematically modelled and a series of possible barrier or enhancement effects can be tested on the model until it is shown to conform to reality. The ultimate purpose of such studies is to advise public health authorities regarding appropriate control measures, even when the mechanisms causing the speed of spread are not fully understood at a medical level.

Geographers working on these topics use their skills in a strictly practical manner likely to lead to considerable saving of public expenditure. Hunter (1966) has demonstrated an ecological approach to the transmission cycle of river blindness in Northern Ghana, where vector and man wage a slow struggle for possession of the fertile riverside locations. As other examples, Brownlea's work (1967; 1972) on the spread through an Australian suburban environment of infectious hepatitis and Tinline's study (1970) of the possible climatic influences which may spread foot-and-mouth virus in Midland England are attacks on diseases which currently cost each country much money and suffering.

Such work upon disease diffusion is the very body of modern quantitative geography with its emphasis upon flows and networks. Indeed, Haggett (1972) has considered measles in South-West England as a planar graph in a recent methodological excursion into a field he once described as 'a confusing sub-variety (of geography)' (1965, p. *v*). His conclusions, although exploratory, stimulate further work in this direction and show that geographers may contribute indirectly to reducing human suffering from disease by furthering our understanding of causative processes.

In brief, the medical geographer's tasks are to prepare and collate disease data and to map them to show where a certain condition is rife (or absent); to apply objective statistical tests to these distributions to assess whether or not the pattern is likely to have occurred by chance; to measure the degree of co-extensiveness between disease and other spatially varying factors; and to apply tests to decide whether any spatial associations he has shown could be causative. He may also need to consider the time/space relationships of disease diffusion processes. Many of his inquiries may come to nothing, but the initiation of even a single new hypothesis will be of value to medicine.

IV Conclusion

The late Sir Dudley Stamp (1964a; 1964b) used to refer to medical geography as 'a tool for research'. With the increasing tendency to modern quantitative analysis in geography, however, it is a tool likely to prove even more useful.

The role of the medical geographer is to make available to medicine the skills of geography but in no way to usurp the functions of medical men. Unfortunately Sir Dudley did not live to see the subject recognized at New Delhi in December 1968 as a full Standing Commission of the International Geographical Union. Although the call for Commission status was supported by most of the member nations, it is the developing countries which probably have more to gain than the developed ones. In terms both of disease distribution and of environmental factors, Birmingham and Sheffield may seem fairly similar. There are great differences, however, between Barotseland and the Zambian Copperbelt. Furthermore, whilst Britain, the U.S.A. and North-West Europe have today an almost monolithic social culture, parts of Africa and Asia vary very greatly from each other. 'In parts of Africa the disease picture resembles that of medieval Europe for the individual may suffer from five or six diseases' (Banks, 1957). In other areas, rising incomes and literacy standards are bringing Africans closer to the public health patterns of Europe (Walker, 1964) and the onrush of development is rapidly doing away with the heterogeneity of the past. This gives real urgency to studies based upon areal differentiation before those differences disappear. In turn this urgency justifies studies in medical geography being attempted without delay. It justifies the acceptance of today's low standards of accuracy and it rejects delay for improvement of diagnoses and statistical data, although these may be possible in the future.

There remains the question of who is to undertake this type of work; but the question itself seems barely relevant, for 'disciplinary boundaries will hopefully remain hazy' (Banta and Fonaroff, 1968). Collaborative effort as co-members of an inter-disciplinary team is likely to yield the best results and even the disciplines represented will vary with the individual problem which requires solution. Most often medical geographers will work with medically qualified men; but it may be in botany or in biochemistry that a particular piece of interpretative scientific search will lie.

Here geographers recognize the need for caution. Medical geography is a tool but rarely an end in itself. It is the application of geographical methods and skills to medical problems. One may consider geographical *evidence* on medical hypotheses. It would be improper to claim that the geography provides *proof*. Geography must inevitably generalize about an area, and thus its hypothetical relationships must also be generalizations. The confirmation needed for such hypotheses will lie with a discipline which, rather than studying the group, studies each individual case.

References

ACLAND, H. W. (1856): *Memoir on the cholera at Oxford, 1854*. London.

ARMSTRONG, R. W. (1964): Environmental factors involved in studying the relationship between soil elements and disease. *American Journal of Public Health* **54**, 1536–44.

(1969): Standardized class intervals and rate computation in statistical maps of mortality. *Annals of the Association of American Geographers* **59**, 382–90.

(1972a): Computers and mapping in medical geography, *in* N. D. MCGLASHAN, editor: *Medical geography techniques and field studies*. London.

(1972b): Medical geography and health planning in the United States, *in* N. D. MCGLASHAN, editor: *Medical geography techniques and field studies*. London.

BANKS, A. L. (1957): Disease and population. *Lancet* **i**, 749–51.

BANTA, J. E. and L. S. FONAROFF (1968): Some considerations in the study of geographic distribution of disease. *Professional Geographer* **21**, 87–92.

BROWNLEA, A. A. (1967): An urban ecology of infectious disease. *Australian Geographer* **10**, 169–87.

(1972): Modelling the geographic epidemiology of infectious hepatitis, *in* N. D. MCGLASHAN, editor: *Medical geography techniques and field studies*. London.

CASE, R. A. M. and J. M. DAVIES (1964): On the use of official statistics in medical research. *Journal of the Institute of Statisticians* **14**, 89–119.

COOK, P. J. (1967): World maps of cancer incidence rates, *in* R. DOLL, editor: *The prevention of cancer: pointers from epidemiology*. Oxford.

COOK, P. J., C. H. COLLIS and J. K. FOREMAN (1971): Cancer of the oesophagus and alcoholic drinks in East Africa. *Mimeo paper of Medical Research Council*. London.

COOK, P. J. and R. O. WALKER (1967): The geographical distribution of dental care in the United Kingdom. *British Dental Journal* **122**, 441–7, 494–9, 551–8.

CHOYNOWSKI, M. (1959): Maps based on probabilities. *Journal of the American Statistical Association* **54**, 385–8.

DEVER, G. E. A. (1972): Leukaemia and housing—An intra-urban analysis, *in* N. D. MCGLASHAN, editor: *Medical geography techniques and field studies*. London.

DE VISE, P., J. LASHOF, R. MORRILL, R. EARICKSON and L. BLOOM (1969): *Slum medicine: Chicago's apartheid health system*. Chicago.

EARICKSON, R. (1970): The spatial behaviour of hospital patients. *University of Chicago, Department of Geography, Research Paper* **124**.

FONAROFF, L. S. (1968): The historical geography of malaria risk in Trinidad, *in* N. D. MCGLASHAN, editor: *Medical geography techniques and field studies*. London.

FORSTER, F. (1966): Use of a demographic base map for the presentation of areal data in epidemiology, *in* N. D. MCGLASHAN, editor: *Medical geography techniques and field studies*. London.

GILBERT, E. W. (1958): Pioneer maps of health and disease in England. *Geographical Journal* **124**, 172–83.

GIRT, J. L. (1971): The location of medical services and disease ecology— some conclusions on the effect of distance on medical consultations in a rural environment. *Mimeo paper of the University of Guelph*.

(1972): Simple chronic bronchitis and urban ecological structure, *in* N. D. MCGLASHAN, editor: *Medical geography techniques and field studies*. London.

GODLUND, S. (1961): Population, regional hospitals, transport facilities and regions. *Lund Studies in Geography, Series B*, **21**.

HAGGETT, P. (1965): *Locational analysis in human geography*. London.

(1972): Contagious processes in a planar graph: An epidemiological application, *in* N. D. MCGLASHAN, editor: *Medical geography techniques and field studies*. London.

HENSCHEN, F. (1962): *The history of diseases*. Stockholm and London.

HIPPOCRATES: *Airs, waters, places*. Translated by W. H. S. JONES (1923). London.

HOPPS, H. C. (1968): *The computerized mapping of disease (MOD) project*. Washington D.C.

HOWE, G. M. (1963, 1970): *National atlas of disease mortality in the United Kingdom*. London.

HUNTER, J. M. (1966): River blindness in Nangodi, Northern Ghana: a hypothesis of cyclical advance and retreat, *in* N. D. MCGLASHAN, editor: *Medical geography techniques and field studies*. London.

JACKMAN, M. E. (1972): Flying doctor services in Zambia, *in* N. D. MCGLASHAN editor: *Medical geography techniques and field studies*. London.

KING, S. H. (1962): *Perceptions of illness and medical practice*. New York.

LEARMONTH, A. T. A. (1972a): Medicine and medical geography, *in* N. D. MCGLASHAN, editor: *Medical geography techniques and field studies*. London.

(1972b): Atlases in medical geography, 1950–1970: A Review, *in* N. D. MCGLASHAN, editor: *Medical geography techniques and field studies*. London.

LEARMONTH, A. T. A. and R. GRAU (1969): *Maps of some standardised mortality ratios for Australia for 1965–6 compared with 1959–63*. Canberra.

LEARMONTH, A. T. A. and G. C. NICHOLS (1965): *Maps of some standaridised mortality ratios for Australia, 1959–63*. Canberra.

MCGLASHAN, N. D. (1965): The scope of medical geography. *South African Geographical Journal* **47**, 35–40.

(1966): Blindness in the Luapula Province of Northern Rhodesia. *Central African Journal of Medicine* **12**, 41–7, 68–73, 86–9.

(1967): Geographical evidence on medical hypotheses. *Tropical and Geographical Medicine* **19**, 333–43.

(1968): The distribution of population and medical facilities in Malawi. *Central African Journal of Medicine* **14**, 249–252.

(1969): Oesophageal cancer and alcoholic spirits in central Africa. *Gut* **10**, 643–50.

Editor (1972): *Medical geography techniques and field studies*. London.

MAY, J. M. (1954): Medical geography, *in* P. E. JAMES and C. F. JONES, editors: *American geography: inventory and prospect*. Syracuse.

(1958): *The ecology of human disease*. New York.

(1961): *Studies in disease ecology*. New York.

MORRILL, R. L. (1966): Hierarchy of hospital services: classification of hospitals. *Chicago Regional Hospital Study Working Paper* **I.8.**

(1967): Relationship between transportation and hospital location and utilization. *Chicago Regional Hospital Study Working Paper* **I.14.**

MORRILL, R. L. and R. EARICKSON (1968): Hospital variation and patient travel distances. *Inquiry* **11**, 1–9.

(1969): Locational efficiency of Chicago hospitals: an experimental model. *Health Research*, Summer.

PALMER, P. E. S. (1963): *Geographical pathology*. An unpublished lecture to the South African Institute for Medical Research, Johannesburg.

PETERMANN, A. (1852): *Cholera map of the British Isles showing the districts attacked in 1831, 1832 and 1833*. London.

PROTHERO, R. M. (1965): *Migrants and malaria*. London.

(1968): Public health, pastoralism and politics in the horn of Africa, *in* N. D. MCGLASHAN, editor: *Medical geography techniques and field studies*. London.

(1969): North-east Africa: a pattern of conflict, *in* R. M. PROTHERO, editor: *A geography of Africa: regional essays on fundamental characteristics, issues and problems*. London.

PYLE, G. F. (1971): Heart disease, cancer and stroke. *University of Chicago, Department of Geography, Research Paper* **134**.

PYLE, G. F. and P. H. REES (1973): Problems of modelling disease patterns in urban areas: the Chicago example. *Economic Geography* [in press].

ROSE, E. F. (1965): Interim report on the survey of cancer of the oesophagus in the Transkei. *South African Medical Journal* **39**, 1098–101.

SAKAMOTO, M. M. and K. KATAYAMA (1966, 1967): A medico-climatological study in the seasonal variation of mortality in the U.S.A. *Papers in Meteorology and Physics* **17**, 279 and **18**, 209.

SNOW, J. (1849, 1855): *On the mode of communication of cholera*. London.

STAMP, SIR L. D. (1964a): *Some aspects of medical geography*. London.

(1964b): *The geography of life and death*. London.

TINLINE, R. (1970): Lee wave hypothesis for the initial pattern of spread during the 1967–8 foot and mouth epizootic, *in* N. D. MCGLASHAN, editor: *Medical geography techniques and field studies*. London.

WALKER, A. R. P. (1964): Some thoughts on the future health of the South African Bantu. *South African Medical Journal* **38**, 255–60.

WARREN, H. V. (1959): Soil and multiple sclerosis. *Nature* **184**, 561.

(1964): Geology, trace elements and epidemiology. *Geographical Journal* **130**, 525–8.

WHITE, R. R. (1972): Probability maps of leukaemia mortalities in England and Wales, *in* N. D. MCGLASHAN, editor: *Medical geography techniques and field studies*. London.

10 Recreation

J. A. Patmore

I Geography and leisure

The possession and enjoyment of leisure is no new phenomenon. To the Greeks, work (*ascholia*) was simply the absence of leisure (*schole*). The state of leisure might in practice be the preserve and the prerogative of an elite, but it was seen as a conscious and desirable goal. For long, however, perception and attainment were far removed. The leisure of the few was founded on the labour of the many and respite from toil was but transient relief. Now, in the developed world, the privilege of the minority has become the possession of the majority, not yet in as abundant or as universal a form as many might desire but still in a hitherto unimagined range of quantity and quality.

Leisure and pleasure are not, of course, synonymous. As the Outdoor Recreation Resources Review Commission of the U.S.A. (1962) was quick to point out, 'leisure is the blessing and could be the curse of a progressive, successful civilization'. For some the possession of abundant leisure may be a fundamental problem, not least when matched by the impoverishment of unemployment or the bodily handicaps of age. For others, a bleak environment in both a social and a physical sense may bring little pleasure and less quality to their leisure experience. But the problems of leisure should not gainsay its prospects: for many, the quality of leisure is one measure at least of the quality of life.

The study of leisure ranges over many disciplines, and in recent years the rise of academic interest has seemingly far outpaced the growth of leisure pursuits themselves (Palmer, 1967; Burton and Noad, 1968). Some of this interest has remained within traditional subject bounds, the obvious and evident concern of such disciplines as sociology, economics, landscape architecture, botany or physical education. But far more has been consciously or unconsciously of genuinely interdisciplinary concern, stimulated by the practical needs of management and planning (Burton, 1970). Teams established by such governmental bodies as the Sports Councils or the Countryside Commissions in Britain have drawn on a wide variety of skills, and a great deal of the resulting work is difficult to assign to a single subject field. Many such distinctions are indeed arbitrary, as other chapters in this volume have clearly shown: while the illumination may arise from a particular point of view, the practical applications permit no narrow focus.

In this sense, therefore, the academic study of leisure has become the apotheosis of applied study with the end rather than the means as the essential goal. Paradoxically perhaps, there is nevertheless a distinctively geographical contribution in this field for, particularly where recreation rather than leisure is concerned, the demand for and the supply of facilities raise problems of an inherently spatial nature, not least in those countries like Britain with varied and urgent demands on scarce resources of land and a necessary ordering of priorities. It is the purpose of this chapter to explore some of the contributions which have been made to these problems from a geographical point of view—though it must be emphasized that the concern is with a point of view rather than the narrower context of the work of geographers as such. As many of the bibliographical references at the end of this chapter clearly indicate, numerous worthwhile studies have been undertaken by researchers who would not consider themselves in any sense academic geographers but who have none the less used inherently geographical techniques.

In parenthesis, a note of definition is necessary. Leisure itself has many meanings (Cosgrove and Jackson, 1972), but the rather negative view adopted by the Countryside Commission (1970a) is particularly apposite in the present context. This regards leisure as 'the time available when the disciplines of work, sleep and other basic needs have been met'. The concern in this essay is more positive, with recreation rather than with leisure, with the positive use of leisure time in a wide variety of pursuits and in particular with outdoor recreation beyond the confines of home and garden, for it is in this context that spatial organization and spatial concerns become paramount.

II Evolution

While the problems posed by recreation are far from new, their scale most certainly is. The basic constraints to participation in recreation are threefold: first, *desire*, the very idea that outdoor recreation can yield a pleasurable experience, and an experience not restricted to the socially and financially privileged; secondly, the *ability* to match desire with adequate leisure time and with adequate disposable income to purchase travel, equipment and accommodation; and lastly ability enhanced by *mobility*, the widespread availability of speedy personal transport.

Historically, three phases may be distinguished in the changing impact of these constraints. Initially, all factors operated to keep demand limited and its satisfaction for the great majority essentially local. A nation of rural dwellers saw little attraction in the rural scene as such, while towns were too small in area for their denizens to feel any real sense of isolation from the surrounding countryside. Recreations were simple in form, making little demand on space (Pimlott, 1968). But there were early seeds of later change. For the privileged few, a static life even amid the pleasures of London society could soon pall. The spas might offer 'but London life on another stage' as Elizabeth Montagu wrote in 1754, but the change of environment they provided was as important as the pleasures they could afford (Patmore, 1968).

In the nineteenth century some elements at least of the privileges of the few came within the grasp of the many. Urban growth was accompanied by a drastic lowering in the quality of the urban environment, but concomitant industrialization brought not only a gradual increase in the wealth of the individual but ultimately also in the time he could call his own. In recreational terms there was a twofold response. Space to promenade and to play had local expression within or adjacent to the urban fabric in the creation of the great open tracts of the Victorian parks (Chadwick, 1966). Beyond the confines of the town, the advent of the railway conferred a new mobility, and the emergence of popular seaside resorts exploited a new resource with almost unlimited capacity (Gilbert, 1939, 1954; Pimlott, 1947). Ironically perhaps, the urban dweller's recreation was still in an urban setting, his perspective those of park, promenade or pier. But there were already substantial hints of change. The ageing Wordsworth, objecting strongly to the proposal for a railway to Windermere, might aver that 'a vivid perception of romantic scenery is neither inherent in mankind nor a necessary consequence of even a comprehensive education', but his beloved Lakes rapidly became a popular destination for the excursion hordes. 'In nineteenth-century Ambleside', wrote James Payne, 'our inns are filled to bursting . . . A great steam monster ploughs up our lake and disgorges multitudes upon the pier; the excursion trains bring thousands of curious, vulgar people . . . our hills are darkened by swarms of tourists; our lawns are picknicked upon twenty at a time, and our trees branded with initial letters . . .' (Margetson, 1969). Congestion and conflict of use were early themes in recreation.

The fundamental changes, however, belong to the present century with escalating desire and ability matched by enhanced mobility. The expression of recreation is no longer concentrated in location but is as widespread in rural as in urban settings, the outward surge of Dower's 'fourth wave' (1965). This new impact of urban man on the rural scene is, above all, a product of the private car. Even by 1939 there were two million cars in Britain: by 1972 there were over twelve and more than one household in two had a car available.

This brief historical diversion is important not only for the trends it reveals, but as a necessary reminder of the fickleness of fashion and the rapidity of change. It is salutary to recall that the principal leisure pursuit today, the watching of television (Sillitoe, 1969), has only been available anywhere in the world as public transmissions since 1936. Rural recreations may have more extended antecedents, but their intensity of impact, and the problems they engender, are recent indeed. Such rapidity of change makes prediction in the leisure field more than usually hazardous (Rodgers, 1969) but the need for accurate analysis and the establishment of effective guidelines for management become increasingly important. In the remainder of this essay, the geographical approach to analysis and management will be more fully explored.

III Demand

Studies in recreation conveniently divide into those concerned primarily with the *demand* for varied pursuits, and those whose focus is the *supply* of available resources. In some senses, the distinction is artificial and unfortunate, for the essence of most problems remains the critical nature of the relationship between demand and supply; but the division remains convenient in any review of approaches and techniques.

Demand studies fall into three broad categories. The first examines the whole pattern of demand of the total population over the full range of leisure activities, a necessary preliminary to any realistic assessment of priorities. Overall studies of demand, however, have the inherent disadvantage that many minority activities are represented by only a tiny proportion of the total population, and in a general survey such small sub-samples may lack statistical significance. In consequence, a second approach seeks to isolate the population pursuing a particular activity and thus to examine its characteristics in far more detail than would otherwise be possible. The third category changes the focus from participant to place and concentrates attention on the site where a specific type of recreation is undertaken.

1   The broad-scale demand pattern

Large-scale demand studies of recreation had their effective origins in the United States. In 1958, Congress established the Outdoor Recreation Resources Review Commission. In addition to its final report (1962), the Commission sponsored a series of 27 study reports which covered a whole range of fundamental topics in the field of recreation and which remain a vital source. In the present context, the most important was Study Report 19, the National Recreation Survey, which reported the results of a nationwide survey of the outdoor recreation habits and preferences of Americans aged 12 and over. The data were derived from four separate samples, each involving some 4,000 interviews. This work gave valuable insights and established important methodological precedents, but the results are obviously not applicable in any detail outside the American context.

There have been two comparable studies in Britain, and it is typical of the interdisciplinary nature of the study of recreation that one was the responsibility of the Government Social Survey (Sillitoe, 1969) and the other of a geographer, H. B. Rodgers (British Travel Association/University of Keele, 1967; 1969), though the basic technique of home-interview questionnaires was the same in each case. It is beyond the scope of this essay to discuss the results of these surveys in detail (Rodgers, 1968, 1969; Patmore, 1970), but taken together they outline the general patterns of recreation at a national level with reasonable confidence. Their prime limitation is the obvious one of a sample too small to permit more detailed consideration of minority activities or of varying demand at a regional scale, a limitation which stems inevitably from the inexorable control of relatively limited finance.

The Pilot National Recreation Survey, in its second report, did attempt at least some analysis at regional level, with a geographer's recognition that 'there is no stereotyped national pattern in our use of spare-time, only a set of complexly varying regional patterns, strongly idiosyncratic, and themselves probably composed of even more intricate sub-regional and local variations'. But, with a national sample of 3,167 respondents, only tantalizing glimpses of regional patterns could be obtained, with a coarse grouping of data into four regions for England, together with Scotland and Wales. Nevertheless, even at this scale, a clearer picture begins to emerge of the close relationships between demand and supply. In the North, for example, hill-walking, climbing and hiking showed a very strong overdevelopment compared with the national picture, surely a clear reflection of the good walking and climbing country within easy reach of many northern industrial towns. Equally, the poor development of sea-sailing in the region was a pointed reminder not only of below-average incomes but of the physical unsuitability for this pursuit of many northern coasts.

These glimpses have since been refined in two instances by full surveys of demand at a regional level (North Regional Planning Committee, 1969; North West Sports Council, 1972). While to the planner the overall regional picture these surveys reveal is of obvious value, to the geographer it is the insight they afford into sub-regional variations which has a particular fascination, revealing still more clearly the fundamental role of spatial variation in resources in influencing both levels and patterns of participation in recreation. One example, from the North West, must suffice. The data reveal that, though the denizens of Merseyside are more likely than those of Manchester to make outings to coast and countryside and to make them more frequently, the distances they travel on such outings tend to be much less. When a full day is available for such an outing, there is little overall contrast in the type of destination sought, but when time is more restricted (Fig. 10.1) fundamental differences emerge: coastal areas predominate for Merseysiders, but town parks play a far more significant role in the case of Manchester. Such variations cannot be related to any social or economic differences between the conurbations: they are rooted rather in the varying opportunities afforded by location. Merseyside is in close proximity to the sandy beaches of South Lancashire and the pleasant countryside of Wirral: Manchester is more amorphous and sprawling in form, its centre further removed from open country and the nearest coast over 55 km away. Far too little is yet known of such local distinctions of recreational habits and their relationships to inequalities of local opportunity, but results such as these certainly suggest that much latent demand may lie concealed in areas with restricted facilities.

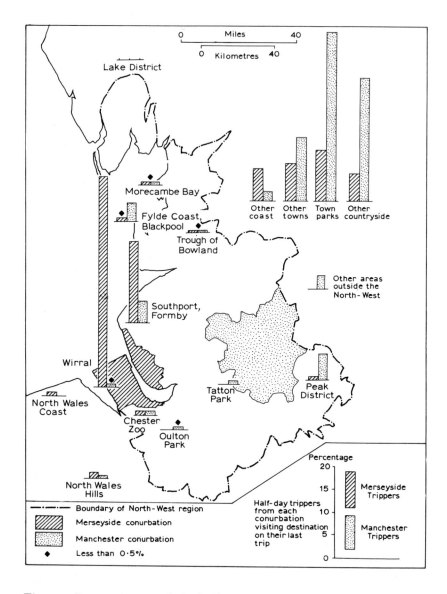

Fig. 10.1 Destinations sought by half-day trippers from the Merseyside and Manchester conurbations. It should be noted that the figures for each destination are expressed as a percentage of all half-day trippers from each of the respective conurbations, and therefore give no indication of the actual numbers travelling to a given destination from the two conurbations. Based on data in North West Sports Council (1972).

2   Demand for particular activities

Even the regional surveys, however, fail to give an adequate picture of many minority pursuits. The general dominance of informal activities is clear enough. In the North West survey, for example, it was shown that 71 per cent of the sample had been on at least one full-day trip in the previous twelve months, but only 37 per cent had taken part in any form of sport in the same period. When individual sports are considered, the totals fall dramatically: only swimming (15 per cent) is well represented, followed by soccer and tennis (5 per cent) and table tennis, bowls and fishing (4 per cent). Though such figures need some care in interpretation, they point to the very real difficulties in obtaining significant data for participants in many activities. In part, this deficit has been met by specific studies of individual pursuits at national or at regional level (for example, Hartley and Knapp, 1970; British Tourist Authority, 1970; County of Denbigh, 1971; Natural Environment Research Council, 1971), but the quality of this research is very far from uniform and major gaps in knowledge still remain.

Given the sporadic nature of the data, it is scarcely surprising that there is little precise knowledge of the actual growth rates of individual activities. Most time series data relate to such material as club membership totals or club affiliations to national organizations and need great care in their interpretation (Molyneux, 1968). While the general trends are clear enough, and in particular the growing emphasis on informal, countryside activities and on water sports (Nicholls and Young, 1968), accurate predictions, even without allowance for the impact of technical innovation and the dictates of fashion, are not readily available, and there have been few attempts to develop an effective methodology (North West Sports Council, 1972).

3   Demand at individual sites

It may with some justification be argued that the expression of an individual's demand for recreation is not wholly a geographer's concern, though much of the work has been undertaken by geographers and many of the research techniques employed have been drawn from the field of social geography. The spatial expression of that demand, however, in the use made of specific recreation sites, is an inherently geographical problem. Site studies have a particular attraction for research. They avoid the difficulties of trying to isolate participants in a specific activity from the general population, though conversely they cannot of course be used to examine the relationship of such participants to the population as a whole. Care must obviously be taken in the interpretation of results. In 1970, for example, the Goyt valley in the Peak District was the scene of an experiment in creating motorless zones in scenically attractive rural areas. Valley roads were closed to traffic at summer weekends: visitors parked their cars at the periphery of the motorless zone and a free minibus service was provided to take them into the valley. The experiment was monitored by a questionnaire survey of the attitudes of visitors to the valley: a satisfying 94 per cent 'supported the experiment in principle', but such results, of course, give no indication of the attitudes of those opposed in principle to the imposition of motorless zones and who in consequence avoided any visit to the Goyt valley at such times (Peak Park Planning Board, 1971).

Surveys have been conducted for a wide variety of sites, ranging in scale from studies of formal facilities in restricted areas such as indoor sports halls (Perrin, 1971) and urban recreation grounds (Winterbottom, 1970) to attempted overviews of informal activities in large tracts of rural terrain (for example, West Riding County Council, 1969; Duffield and Owen, 1970; 1971; Hennessy and Mansfield, 1970; Peak Park Planning Board, 1970; Board, et al., 1972). Much of the initial attraction of site surveys lies in the worthwhile results which can be achieved by a single individual or a small team with restricted resources of finance and time. Burton's classic study of Box Hill (1966) was the effective progenitor of a large number of surveys of informal recreation areas, whose results have been summarized by Patmore (1970) and Houghton-Evans and Miles (1970a). While each site possesses its own distinctive characteristics, the accumulation of evidence from widely differing areas enables some generalizations to be made with confidence and the lessons of such generalizations to be applied in management proposals both for single facilities and for areas as extensive, and as sensitive to recreational pressures, as the New Forest (New Forest Joint Steering Committee, 1970; 1971).

It must nevertheless be admitted that there are many obstacles to effective generalizations. Most early surveys, whether carried out by individual workers or by county planning departments, were conceived in isolation to meet the demands of a particular situation. The consequence has been an inevitable but frustrating lack of comparability, whether in the nature of the sample chosen, the questions asked or the definitions used. The Countryside Commission (1970b; 1970c) has sought to introduce a measure of standardization in such work, though too late to influence the design of many substantial studies. Further problems necessarily arise when consideration is extended to more than a single site, or where the survey is continued over a long period of time. The availability of adequate finance and labour to conduct such large-scale surveys is obviously restricted, though there have been instances involving the simultaneous use of ground and aerial observers to permit an effective picture of activities at county scale to be obtained (Furmidge, 1969).

The lessons of site surveys range over many topics. The varying intensity of use through time, in annual, weekly or daily rhythms, is clearly important when priorities for the investment of scarce capital resources are considered. While the basic constraints on such periodicity are obvious enough, some have received less than adequate attention. Notable among these is the impact of weather in affecting not only the number of visitors to coast and countryside on a given occasion, but also the specific type of destinations chosen. Wager (1967) in his work on commons devised a rough index of attendance at open spaces under different weather conditions. More recently, Bruning (Perry, 1972) has concluded that, for the Netherlands at least, temperature probably exerts the strongest climatic effect on visitor totals, independent of actual insolation, and may be used to some extent at least to predict likely numbers on a particular day.

To the geographer, patterns of use are fundamental. This spatial concern is extended in two directions. In the first place there are the patterns of movement to a site, the attraction exerted by its location. Secondly, there are the patterns of activity generated at the site itself. Knowledge of both types of pattern arises from the study of expressed demand, but in its application profoundly influences the effectiveness of supply.

(a) *The attraction of sites.* The journey to a recreational site is a fundamental part of the whole recreational experience (Clawson and Knetsch, 1966). Indeed, 'going for a drive' may be a recreational end in itself, difficult to monitor in that it involves no use of actual recreational sites in the accepted sense of the phrase (Colenutt, 1969). Patterns of movement may themselves change rapidly, through the more ready availability of personal transport or through improved access brought about by such technical advances as railway electrification or motorway construction (Jackson, 1970).

Patterns of movement are obviously related to the distribution pattern of recreational facilities; yet the location of recreation sites in relation to the direction and intensity of demand has rarely received the attention it deserves. The fact that the scale of demand is in part conditioned by opportunity has already been demonstrated (Fig. 10.1), but ideas on what constitutes 'effective location' remain rudimentary. Nowhere is this more true than for open space within the urban fabric. In Britain, much urban open space provision remains the seemingly haphazard legacy of historical accident (Chadwick, 1966), but even the guidelines to appropriate levels of provision which do exist are based all too often on 'crude assumptions and arbitrary assertions in place of established facts' (Liverpool Corporation, 1964). It is almost incredible that for so long the only real guidance available was the National Playing Fields Association's standard of 2·4 ha (6 acres) of permanently preserved playing space per thousand population, first formulated in 1925 and reaffirmed as recently as 1969 (National Playing Fields Association, 1971). The assumptions on which this standard was based were arbitrary in the extreme, yet it was commended as a planning guide by successive governments, with the addition of 0·4 ha of ornamental public open space per thousand population. Even so, the standard is concerned only with *levels* of provision, not its *location*. In the latter regard, the N.P.F.A. could only assert that

> The provision of an adequate *total* area of playing space is not, on its own, the complete answer. It is also important to see that the playing space is situated where it is most wanted. If it is so far away from the centres of population, or is so inaccessible that it is not used, the purpose of its provision is largely defeated (p. 9).

Nor is the N.P.F.A. the only body to rely on exhortation and the most general of suggestions in this respect. Even a recent Sports Council handbook can do no more in respect of playing areas than affirm: 'all we can say is that they should be readily accessible' (Sports Council, 1968).

More precise ideas can be obtained by delineating the catchment areas of existing facilities through site surveys (Greater London Council Planning Department, 1968). One such survey in Liverpool (Balmer, 1973) studied 41 of the city's 96 parks and completed questionnaires with over 13,500 park users. Fig. 10.2 illustrates in simple form the results obtained. When all data are aggregated (Fig. 10.2A), the dominance of short-distance movement is evident. More than half of all park visitors (52·5 per cent) travel less than a quarter of a mile and virtually 80 per cent less than three-quarters of a mile. These broad conclusions are obviously capable of much refinement. Association analysis reveals four clear groupings (Fig. 10.2B), with differing types of park having varying sizes of effective catchment area. These park types are not related to size of park alone, but to the quality and range of facilities available. Using the results of this survey, a threefold hierarchy of park types has been suggested as best able to fulfil the outdoor recreation needs of the city dweller,

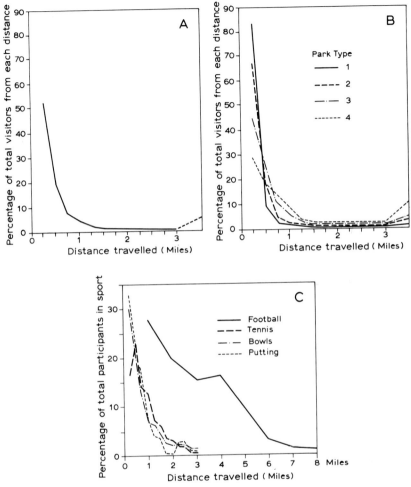

Fig. 10.2 Distances travelled to open spaces in Liverpool. (A) distances travelled to all parks; (B) distances travelled to differing types of park; (C) distances travelled to sports facilities. Data from Balmer (1973).

ranging from a 'neighbourhood' park of 0·8–2·0 ha, with an effective catchment area of 0·4 km, through a 'local' park of 14·2–18·2 ha and a catchment area of 0·8 km, to a 'city' park of over 40 ha and with a city-wide catchment area.

The objective determination of locational criteria is not confined to urban sites alone. Indeed, the distinction usually drawn between 'urban' and 'rural' provision for outdoor recreation is a false one: to most urban dwellers, the perceived distinction is rather one of 'local' and 'distant' with opportunities for use conditioned by time, distance and cost. The 'city' park of Balmer's study could equally well lie in the urban periphery, 'green belt' rather than 'urban' in location, perhaps a 'country park' within the meaning of the 1968 Countryside Act where one indication of need is a 'deficiency of recreation areas within easy reach of large city regional populations'.

Most site studies have included as a fundamental component some information about catchment areas, and even in the countryside the general picture emerges from this evidence of dominantly short distance recreational movement (Law, 1967). Once again, a hierarchy of sites may be recognized, ranging from those whose resources are unique by any standard and whose appeal is national to those of lesser quality and of purely local significance. Law envisages five levels in the hierarchy, with the levels distinguished by extent of catchment area as much as by quality of resource.

(b) *Patterns at a site.* While the location of a site, the patterns of movement to it and its consequent place in a hierarchy of provision are fundamentally geographical concerns, so too are the actual patterns of activity at the site itself. At the micro-level, there have been numerous studies of the observed behaviour of participants at a wide variety of sites (for example, Burton, 1966; Lindsey Countryside Recreational Survey, 1967; Wager, 1967; Mutch, 1968; Furmidge, 1969; Peak Park Planning Board, 1970; Board *et al.*, 1972): these underline the general picture of informal, passive and frequently car-oriented pursuits. But more important at the macro-scale is the general pattern of the impact of urban man on the rural scene. It must be recognized that demand is not blanket in its expression, but rather is confined to narrow corridors of movement and to nodes in the network of routes, to recreational gathering points, or 'honeypots' in the contemporary jargon.

The role of the car in outdoor recreation has already been underlined, and its dominance naturally focuses attention on the network of roads. But even for those who leave cars behind, the use of land is rarely general in its impact, except at such nodal sites as picnic areas, and parks and sports grounds reserved specifically for recreational use. The rural scene is observed from the corridors of roads or footpaths: its visual quality is obviously important, but actual access much less so. Even on open tracts of upland, most public movement is confined to existing paths. In the Peak District, only 5 per cent of visitors were observed to leave the paths on grouse moors despite access agreements (Picozzi, 1971): similarly in the Lake District, 96 per cent of visitors 'always' or 'nearly always' kept to the defined path beside Stickle Ghyll in Great Langdale (Gee, 1969).

On a regional scale, the concept of a crowded countryside is but relative at best (Fig. 10.3). The pressure on well-loved nodes may indeed be intense, but such nodes still form only a tiny minority of potential destinations. In Snowdonia, certain sections of the coast, the Snowdon range itself, Nant Ffrancon and Lake Bala are truly crowded, but much of the remainder is only lightly used and substantial parts are as yet scarcely disturbed by recreational activity. The recognition and the delineation of these patterns, an inherently geographical exercise, is the essential prerequisite for the effective management of such areas, with its emphasis on the horizontal segregation of activities (for example, Peak Park Planning Board, 1972).

IV Supply

The areal expression of patterns of demand inevitably trespasses on a consideration of the supply of recreational resources. Already, in the study of demand, the author may with justification be accused of a parochial point of view, of an undue focus on British examples of practice if not precept. Recreational patterns, however, inevitably have a distinctive national flavour, conditioned not only by factors of social tradition, standards of living and climate, but also by the fundamental space relationships of land and population. This is even more true of the supply side of the equation, where the overall availability of land, and the location of such land in relation to the major concentrations of population, are of crucial importance.

A brief contrast may be drawn in this respect between British and American conditions. In the U.S.A., over one-third of the land remains in Federal ownership, and 114,000,000 ha, or almost one-eighth of all land, are classified as recreation areas, that is publicly owned and managed land on which recreation is a recognized use. With this abundance, despite severe regional imbalances, it is scarcely surprising that most American outdoor recreation takes place on areas designated for that specific purpose. Indeed, of the six types of recreation resource distinguished by the Outdoor Recreation Resources Review Commission (1962), on only one, Class III, natural environment areas, was recreation envisaged as being 'usually in combination with other uses'. In Britain, in contrast, little land is managed exclusively for recreation. It has been estimated that in England and Wales some 1,210,000 ha are in recreational use of some kind (Burton and Wibberley, 1965), but most of this area comprises common land and woodland where recreation may play a very

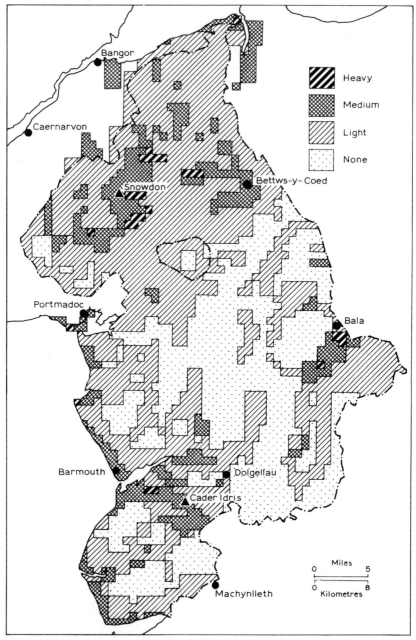

Fig. 10.3 The intensity of the recreational use of the Snowdonia National
Park. Intensity of use is measured on a four-point scale for each km² of the
national grid. Data from J. W. E. H. Gittins (unpublished report for the
Snowdonia National Park Survey).

subordinate role. Problems of multiple use, and of access to agricultural land, are dominant, problems of only minor relevance in the American context. Much of the discussion of supply must therefore relate explicitly to the British situation.

1   *Inventories of supply*

Simple inventories of resources are bedevilled by difficulties of definition and recognition. So far as sports facilities are concerned, an attempt was made in the 'initial appraisals' of the nine regional sports councils in England and of the Sports Councils for Wales and for Scotland to assess provision, though these appraisals lamentably lacked a uniform brief. Some of their findings have now been brought together at national level, but as yet only so far as they relate to swimming pools, indoor sports centres and golf courses (Sports Council, 1972). On a more wide-ranging basis, assessments have been made of the availability and distribution of recreational resources by Coppock (1966) for Britain as a whole and by Patmore (1970) for England and Wales.

Inventories as such usually give little measure of the quality of resources, though as has already been noted much of the pleasure of recreation in rural areas is rooted in the visual quality of the rural scene. Assessments of quality may in turn form the basis of planning designations, not only as measures of conservation but as a means for defining areas where access and facilities for public enjoyment need to be provided. It is beyond the scope of this chapter to consider the conservation side of the coin, but the assessment of quality is germane to the recreation theme.

The perception of landscape quality is inherently personal, but a geographer's eye for country can be a tangible asset in this respect. The value of such an eye does not always receive recognition beyond the confines of the subject, but in one instance at least has been used as a basis for planning designation. In 1970, the Countryside Commission sought to identify potential 'Heritage Coasts', outstanding stretches of high scenic quality along the remaining areas of substantially undeveloped coast, which could thereby be protected and enjoyed for all time through effective management. In the words of the report (Countryside Commission, 1970d): ⟨

> Those who have visited many different parts of Britain's coastline will have their own ideas of the most beautiful stretches. Some of the better-known parts are so compellingly beautiful in any weather or season that their inclusion in any national list would be readily accepted. Such a list would probably exclude many superb stretches which are inaccessible except on foot and which will have been visited only by those who

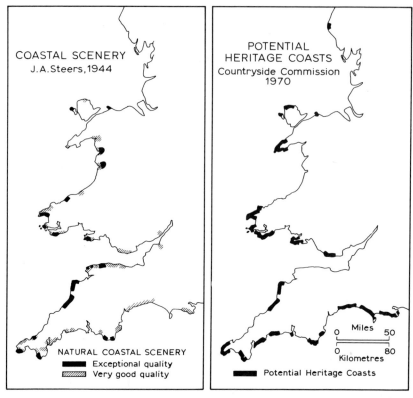

Fig. 10.4 The assessment of the scenic quality of the coast. The work of Steers (1944) was used as the initial basis for the designation by the Countryside Commission (1970d) of potential Heritage Coasts.

have made a special effort to reach them or who have had some special reason for doing so. In order to be comprehensive and include such areas the most acceptable method would be for a skilled observer to examine critically every mile of coast, irrespective of access, and record his impressions (p. 11).

Such an assessment was found in the work of Steers (1944) and on this basis, with amendments occasioned by subsequent development, potential 'Heritage Coasts' were delineated (Fig. 10.4).

It is rare, of course, for one person, and a trained person at that, to be able to view *in toto* the whole of an extensive potential resource and assess its worth objectively with a personal but consistent judgement. There have been several recent attempts to create an objective but more universally applicable method of scenic assessment, attempts which have been summarized in a report for the Countryside Commission for Scotland (1971). But landscape quality is only one element in any assessment of an area from a recreational point of view: recreation potential is compounded of site resources of land and water, and of ecological composition as well as scenic worth. Yet if recreational resources are to be fully recognized, and not least if the claims of recreation against other uses of land and water are to be objectively evaluated, some assessment of the recreational potential of an area is worth attempting. Discussion of an appropriate methodology is beyond the scope of this chapter; but Duffield and Owen (1970), in their study of Lanarkshire, made a notable pioneer assay (Fig. 10.5). Varied attributes were mapped, and the cumulative scores of unit areas formed the basis of the final map of recreational potential.

2 The capacity of sites

The recognition and delineation of recreation resources is only the beginning of an assessment of their potential effectiveness. Fundamental to any inventory or evaluation is some determination of capacity, of the number of people that can be accommodated by a specific resource at a given time. Of all recreational concepts, that of capacity is the most difficult to refine. Any area, indeed, has several capacities from a recreational point of view. *Physical capacity* is the maximum number of people a site can physically accommodate for a given activity. It is rare that any site is used to its physical capacity for any activity, except for spectator sports, for few people would tolerate the degree of crowding this would entail. In some recreational circumstances, physical capacity may be determined not by the site itself, but by the capacity of such ancillary facilities as car parks. Of greater relevance under most conditions is the *psychological* or *social capacity* of an area for recreation, of the number of people it can absorb before the latest arrivals perceive the area to be 'full' and seek satisfaction elsewhere. Capacity in this sense is inevitably a personal affair, compounded of an individual's perception tempered by variations of mood, activity, time and place. Yet, abstract though the concept may seem, some attempt must be made to establish standards, however arbitrary, if any realistic measure of effective capacity is to be made. Not unnaturally, such attempts engender as much heat as light (Houghton-Evans and Miles, 1970b; Kirby, 1971), and workers in differing localities, or with differing points of view, arrive at varying conclusions (Dower, 1966; Weal, 1968; Gee, 1972).

A third concept of capacity is more tangible, that of *ecological capacity*—the level of recreational activity an area can undergo before

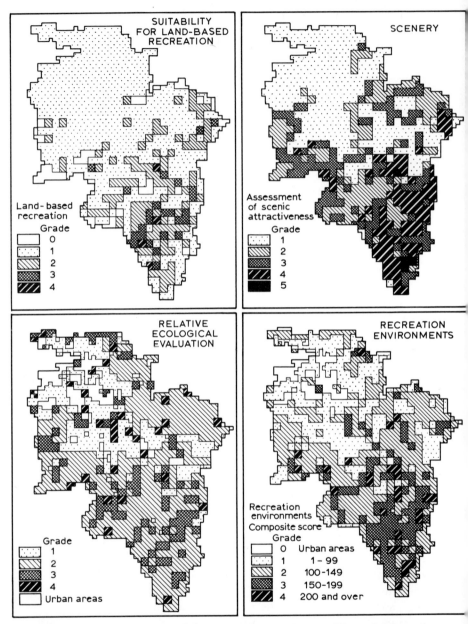

Fig. 10.5 An assessment of the recreation resources of Lanarkshire, after Duffield and Owen (1970). The aim of the technique is to delineate 'recreation environments' (bottom right) as an indication of the varying capacity of the county to support and sustain recreational activities. These recreation environments are synthesized from assessments of suitability for land- and water-based recreation (the latter not illustrated here) and of scenic and ecological importance. The techniques of assessment are discussed in Duffield and Owen (1970), Appendix 2.1. The unit of assessment is normally 2 km × 2 km squares.

irreversible ecological damage is sustained (see also Ch. 11). The basic problems of trampling and resultant wear have long been the subject of research (Bates, 1935; Davies, 1936), but recent increases in recreational pressures have increased the urgency and practical relevance of this work (Duffey, 1967; Lloyd, 1970). Botanists and biogeographers have shared a common ecological concern, and devised techniques not only to analyse the problems of sensitive habitats (Goldsmith, Munton and Warren, 1970; Bayfield, 1971; Chappell, Ainsworth, Cameron and Redfern, 1971; Streeter, 1971) but to suggest management methods whereby the problems can be contained and habitats preserved without undue deprivation of public enjoyment (Huxley, 1970; East Lothian County Council, 1970).

V Demand/supply relationships

Through almost every study of the geography of recreation there runs a practical thread, for that study, as was early emphasized in this chapter, was nurtured in a growing awareness of urgent problems intimately related to the whole quality of life. The expression of recreation is as much spatial as social or economic, and any analysis of the patterns of demand or any assessment of the resources of supply is inevitably rooted in geographical techniques. In many ways, the distinction between demand and supply is arbitrary and academic, as many preceding examples have shown, for the ultimate concern is the relationship between them.

Supply, to be effective, must satisfy as fully as possible in its location the expression of demand. In a situation like that in Britain where demand so frequently outstrips supply, there is a tendency to neglect this locational element, for almost any provision, in any location, will match some need and yield in consequence a visible return on investment: the problem seems to be more often the curbing of demand to a level which can be contained by the available supply. Sensitive areas of high quality resources need active conservation: distance alone from centres of population is an inadequate contraint (Ballantine, 1971). Positive provision has had a low priority for scarce capital resources, and recreational work tends to be an early casualty when the economic winds are chill.

In such a context, a conscious appraisal, at the appropriate national, regional or local scale, of the demand/supply relationship is all the more necessary if a given investment is to meet the greatest need in the best way. Regrettably, much recreational provision has, in the past, been on an *ad hoc* basis even where resources have been readily

available. The most recent example can be seen in the recommendations of grant aid for country parks under the provisions of the 1968
Countryside Act. While the Countryside Commission follows clearly
ordered priorities in assessing whether or not to recommend grant
aid, the responsibility for initiating developments lies firmly with
local authorities or private landowners. The Commission may only
'advise, stimulate and encourage', and in their 1970–71 Annual
Report they had regretfully to record that

> the degree of activity in the provision of country parks, picnic sites and
> other opportunities for informal recreation in the countryside has been
> disappointingly uneven . . . 17 out of 58 administrative counties in
> England and Wales had failed to produce a single scheme for a country
> park or picnic site on which the Commission could recommend grant
> aid . . . Of course, the need for investment is greater in some areas
> than others; but several of the counties which produced no, or exces-
> sively modest, proposals contain popular holiday areas or adjoin large
> cities (pp. 2–3).

More conscious consideration of the location of provision is likely to
arise with the widespread adoption of structure planning and an
active concern at county level to try to match demand and supply
(for example, Lindsey County Council, 1972). It is to be hoped that
the development of the recreational potential of the critically located,
but frequently neglected, urban periphery is not overlooked in the
restructuring of local government responsibilities, and that the false
dichotomy between urban and rural recreation in this respect is not
thereby unhappily preserved (Cracknell, 1967).

Issues of this kind highlight the geographer's role in recreation
studies. This chapter has inevitably been heterogeneous in content,
for recreation geography lacks a single focus or a messianic concern.
Indeed, many issues have been conspicuous by their absence. There
has been no attempt, for example, to assess the contribution of
recreation to regional or local economies (Smith, 1968; Lavery, 1971);
to evaluate recreation in the light of the claims of competing uses for
land (Northern Pennines Rural Development Board, 1970); to
examine in detail the relative potential of differing resources of land
and water for recreation (Tanner, 1969; Appleton, 1970); or to
assess the role of the geographer in devising techniques to interpret
the rural scene for the urban visitor. It must suffice to have indicated
both the reality of geographical concern and the relevance of a geographical approach to analysis, and also to have illumined the role
in recreation of, in John Dower's happy phrase, creative geography.

References

APPLETON, J. H. (1970): *Disused railways in the countryside of England and Wales.* London.

BALLANTINE, G. E. (1971): Planning for remoteness. *Journal of the Town Planning Institute* 57, 60–64.

BALMER, K. R. (1973): *Open space in Liverpool.* Liverpool.

BATES, G. H. (1935): The vegetation of footpaths, sidewalks, cart-tracks and gateways. *Journal of Ecology* 23, 470–87.

BAYFIELD, N. (1971): Some effects of walking and skiing on vegetation at Cairngorm, *in* E. DUFFEY, editor: *The scientific management of animal and plant communities for conservation.* Oxford.

BOARD, C., *et al.* (1972): Leisure and the countryside: the example of the Dartmoor National Park, *in* M. CHISHOLM, editor: *Resources for Britain's future.* London.

BRITISH TOURIST AUTHORITY (1970): *Survey of mobile caravanning and camping 1970.* London.

BRITISH TRAVEL ASSOCIATION/UNIVERSITY OF KEELE (1967): *Pilot National Recreation Survey, Report no. 1.* London.

(1969): *Pilot National Recreation Survey, Report no. 2.* London.

BURTON, T. L. (1966): A day in the country—a survey of leisure activity at Box Hill in Surrey. *Chartered Surveyor* 98.

Editor (1970): *Recreation research and planning, a symposium.* London.

BURTON, T. L. and G. P. WIBBERLEY (1965): *Outdoor recreation in the British countryside.* Ashford.

BURTON, T. L. and P. A. NOAD (1968): *Recreation research methods, a review of recent studies.* Birmingham.

CHADWICK, G. F. (1966): *The park and the town.* London.

CHAPPELL, H. G., J. F. AINSWORTH, R. A. D. CAMERON and M. REDFERN (1971): The effect of trampling on a chalk grassland ecosystem. *Journal of Applied Ecology* 8, 869–83.

CLAWSON, M. and J. L. KNETSCH (1966): *Economics of outdoor recreation.* Baltimore.

COLENUTT, R. J. (1969): Modelling travel patterns of day visitors to the countryside. *Area* 2, 43–7.

COPPOCK, J. T. (1966): The recreational use of land and water in rural Britain. *Tijdschrift voor Econ. en Soc. Geografie* 57, 81–96.

COSGROVE, I. and R. T. JACKSON (1972): *The geography of recreation and leisure.* London.

COUNTRYSIDE COMMISSION (1970a): *Countryside recreation glossary.* London.

(1970b): *Outdoor recreation surveys: the design and use of questionnaires for site surveys.* London.

(1970c): *Outdoor recreation information: suggested classifications for use in questionnaire surveys.* London.

(1970d): *The coastal heritage.* London.

COUNTRYSIDE COMMISSION FOR SCOTLAND (1971): *A planning classification of Scottish landscape resources.* Perth.

COUNTY OF DENBIGH (1971): *Caravanning in Denbighshire.* Denbigh.

CRACKNELL, B. (1967): Accessibility to the countryside as a factor in planning for leisure. *Regional Studies* 1, 147–61.

DAVIES, W. (1936): Vegetation of grass verges and other excessively trodden habitats. *Journal of Ecology* **26,** 38–49.

DOWER, M. (1965): *The challenge of leisure.* London.

(1966): *Planning for amenity and tourism.* Dublin.

DUFFEY, E., editor (1967): *The biotic effects of public pressure on the environment.* Monks Wood.

DUFFIELD, B. S. and M. L. OWEN (1970): *Leisure + Countryside = : A geographical appraisal of countryside recreation in Lanarkshire* (Vol. 1). Edinburgh.

(1971): *Leisure + Countryside = : A geographical appraisal of countryside recreation in the Edinburgh area* (Vol. 2). Edinburgh.

EAST LOTHIAN COUNTY COUNCIL (1970): *Dune Conservation 1970: North Berwick Study Group Report.* Haddington.

FURMIDGE, J. (1969): Planning for recreation in the countryside. *Journal of the Town Planning Institute* **55,** 62–7.

GEE, M. H. (1969): Lakeland footpath survey, *in* BRATHAY EXPLORATION GROUP: *Annual Report* **II.** Ambleside.

(1972): *Lake District footpaths: a study of their capacity.* Ambleside.

GILBERT, E. W. (1939): The growth of inland and seaside health resorts in England. *Scottish Geographical Magazine* **55,** 16–35.

(1954): *Brighton, old ocean's bauble.* London.

GOLDSMITH, E. B., R. J. C. MUNTON and A. WARREN (1970): The impact of recreation on the ecology and amenity of semi-natural areas: methods of investigation used in the Isles of Scilly. *Biological Journal of the Linnaen Society* **2,** 287–306.

GREATER LONDON COUNCIL PLANNING DEPARTMENT (1968): *Surveys of the use of open space* (Vol. 1). London.

HARTLEY, H. M. and B. N. KNAPP (1970): *Riding for recreation in the West Midlands.* London.

HENNESSY, B. and N. W. MANSFIELD (1970): *Pleasure traffic and recreation in the Lake District.* London.

HOUGHTON-EVANS, W. and J. D. MILES (1970a): Weekend recreational motoring in the countryside. *Journal of the Town Planning Institute* **56,** 392–7.

(1970b). Environmental capacity in rural recreation areas. *Journal of the Town Planning Institute* **56,** 423–7.

HUXLEY, T. (1970): *Footpaths in the countryside.* Perth.

JACKSON, R. T. (1970): Motorways and National Parks in Britain. *Area* **4,** 26–29.

KIRBY, C. P. (1971): The recreational capacity of Mid-Wharfedale. *Journal of the Royal Town Planning Institute* **57,** 458–60.

LAVERY, P. (1971): *Patterns of holidaymaking in the Northern Region.* Newcastle upon Tyne.

LAW, S. (1970): Planning for outdoor recreation in the countryside. *Journal of the Town Planning Institute* **53,** 383–6.

LINDSEY COUNTRYSIDE RECREATIONAL SURVEY (1967): *A survey of the public use of the Lincolnshire Wolds and adjacent areas.* Lincoln.

LINDSEY COUNTY COUNCIL (1972): *Lindsey Countryside Recreational Survey.* Lincoln.

LIVERPOOL CORPORATION (1964): Open space. *Review of City Development Plan, Report* **15.**

LLOYD, R. J. (1970): *Countryside recreation: the ecological implications*. Lincoln.
MARGETSON, S. (1969): *Leisure and pleasure in the nineteenth century*. London.
MOLYNEUX, D. D. (1968): Working for recreation. *Journal of the Town Planning Institute* 54, 149–57.
MUTCH, W. E. S. (1968): *Public recreation in National Forests: a factual survey*. London.
NATIONAL PLAYING FIELDS ASSOCIATION (1971): *Outdoor playing space requirements: review of N.P.F.A. playing space target 1971*. London.
NATURAL ENVIRONMENT RESEARCH COUNCIL (1971): *National Angling Survey 1970*. London.
NEW FOREST JOINT STEERING COMMITTEE (1970): *Conservation of the New Forest*. Lyndhurst.
(1971): *Conservation of the New Forest: final recommendations*. Lyndhurst.
NICHOLLS, D. C. and A. YOUNG (1968): *A report on recreation and tourism in the Loch Lomond area*. Glasgow.
NORTHERN PENNINES RURAL DEVELOPMENT BOARD (1970): *The changing uplands*. Appleby.
NORTH REGIONAL PLANNING COMMITTEE (1969): *Outdoor leisure activities in the Northern Region*. Newcastle.
NORTH WEST SPORTS COUNCIL (1972): *Leisure in the North West*. Salford.
OUTDOOR RECREATION RESOURCES REVIEW COMMISSION (1962): *Outdoor recreation for America*. Washington.
PALMER, J. E. (1967): Recreational planning—a bibliographical review. *Planning Outlook* 2, 19–69.
PATMORE, J. A. (1968): The spa towns of Britain, *in* R. P. BECKINSALE and J. M. HOUSTON, editors: *Urbanization and its problems*. Oxford.
(1970): *Land and leisure*. Newton Abbot.
PEAK PARK PLANNING BOARD (1970): *Peak District National Park Visitor Surveys 1969*. Bakewell.
(1971): *Nineteenth annual report of the Planning Board*. Bakewell.
(1972): *Routes for people, an environmental approach to rural highway planning*. Bakewell.
PERRIN, G. A. (1971): *Community sports halls*. London.
PERRY, N. (1972): Recreation activity in the Netherlands: four recent use studies. *Recreation News Supplement* 6, 2–5.
PICOZZI, N. (1971): *Breeding performance and shooting bags of red grouse in relation to public access in the Peak District National Park, England*. Bakewell.
PIMLOTT, J. A. R. (1947): *The Englishman's Holiday*. London.
(1968): *Recreations*. London.
RODGERS, H. B. (1968): Leisure in adult life. *Royal Society of Health Journal* 88, 190–94.
(1969): Leisure and recreation. *Urban Studies* 6, 368–84.
SILLITOE, K. K. (1969): *Planning for leisure*. London.
SMITH, R. J. (1968): *The measurement of the economic benefits of recreation: a critical survey of the literature and of the development of the theory*. Birmingham.
SPORTS COUNCIL (1968): *Planning for sport*. London.
(1972): *Provision for sport*. London.
STEERS, J. A. (1944): Coastal preservation and planning. *The Geographical Journal* 104, 7–27.
STREETER, D. T. (1971): The effects of public pressure on the vegetation of

chalk downland at Box Hill, Surrey, *in* E. DUFFEY, editor: *The scientific management of animal and plant communities for conservation.* Oxford.

TANNER, M. F. (1969): Coastal recreation in England and Wales, *in* COUNTRYSIDE COMMISSION: *Coastal recreation and holidays.* London.

WAGER, J. (1967): Outdoor recreation on common land. *Journal of the Town Planning Institute* **53**, 398–403.

WEAL, F. (1968): Traffic in the country—a strategy for the protection of the Lake District. *Architect's Journal* **98**.

WEST RIDING COUNTY COUNCIL (1969): *Recreation traffic in the Yorkshire Dales.* Wakefield.

WINTERBOTTOM, D. M. (1970): Planning for outdoor recreation in urban areas. *in* T. L. BURTON, editor: *Recreation research and planning.* London.

11 Conservation

I. G. Simmons

I MAN AND ENVIRONMENT—1 *The history of man/environment relations as a theme for study.* (*a*) Pre-nineteenth century times. (*b*) The nineteenth century and after. 2 *Ecological studies of man/environment relations.* (*a*) The development of ecological views by biologists. (*b*) The contribution of geographers. (*c*) Ecology, environment and conservation. 3 *Economics and environment.* 4 *Ethology: the behavioural approach to environment attitudes.* (*a*) The individual. (*b*) The group. 5 *Ethics: normative behaviour towards environment.* (*a*) Historical development of ethical attitudes. (*b*) Time present and time future.

II CONSERVATION AT REGIONAL AND NATIONAL SCALES—1 *Conservation as preservation.* 2 *Conservation as wise use.* 3 *Conservation as environmental management.*

III CONSERVATION AT THE GLOBAL SCALE—1 *The ecological basis of the space-ship earth concept.* (*a*) A stable human population. (*b*) Research on ecosystem stability. (*c*) Pollution control. (*d*) Cycling of materials. 2 *Economic adjustments on the space-ship earth.* (*a*) Consumption patterns. 3 *The motivation for advocating the space-ship earth model.* 4 *Implementation.*

IV INTERACTION OF GEOGRAPHY AND CONSERVATION. 1 *Local and national scales.* 2 *The global scale.* 3 *Values.*

In the writer's undergraduate days, before the advent of the methodological thought-police, it was still permissible to say that 'geography is what geographers do'. In the same way, conservation has for long been a body of ideas and actions, circumscribed only by what conservationists thought and did. Yet for an essay like this such vaguenesses are clearly unsuitable, and it is necessary to be clear about the limits within which it is to move. Just as there have been distinct traditions of study in geography, so have there been in conservation; again, just as all the geographical traditions have in common the examination of phenomena in a spatial context, so the conservation themes share a concern over the relationship between man, the materials of the planet which he uses, and the environments which he inhabits. Because of man's status as a cultural animal, conservation, or the lack of it, results from an attitude towards his 'environment', where environment is taken to mean not only the congeries of physical phenomena that comprise the planet, but also the particular subset which he perceives to be useful to him, namely resources. And because of the spatial ubiquity, numerical strength and technological outreach of *Homo sapiens*, the concern of conservation is that of the relationship between population, resources and environment.

This essay will attempt to explore that theme by concentrating largely upon conservation itself; it does not attempt to involve the geographer, except in passing in section II, until the closing paragraphs, by which time it is hoped that discerning readers will have recognized the fields in which geography has a contribution to make. The four parts of this essay deal with the lineaments of the study of the relations of man and environment (section I) and upon the ways in which the ideas of conservation have worked out in practice at various scales (section II). The third will then examine a radical alternative to most of the ideas previously presented and applicable at the global scale. Section IV is concerned with the contribution of contemporary geographical thought not only to the conceptual development of conservation but also to its practical application. For here lies the justification for such a chapter in this book: few people would dispute that man's use of the planet is producing effects which are not only culturally unacceptable but may possibly lead to such ecological instability that his very survival as a species is threatened. If geographers can help to avert such a condition then, according to the conventional wisdom at any rate, they will be in line for an award of Hero of the Biosphere.

I Man and environment

The present time cannot arrogate to itself a monopoly of thought upon this topic. Throughout literate history there have been opinions expressed upon the world of nature and of man's role within and outside it. Contemporary views can be classified into the four 'e's—ecological, economic, ethological and ethical. Some of these intellectual approaches, which have their roots deep in Western intellectual history, have become incorporated into geographical modes of thought (Prince, 1971), others have so far stayed apart from geography as it is now known.

1   *The history of man/environment relations as a theme for study*

The written deliberations of all manner of men upon this theme from the times of classical antiquity until the eighteenth century have been reviewed exhaustively by Glacken (1967), and in other contributions (Glacken, 1966; 1970) he has reviewed some of the post-eighteenth century developments of thought.

(*a*) *Pre-nineteenth-century times.* From the immense diversity of recorded thought, a number of crystallizing abstractions can be picked out which emphasize the main attitudes to the relationship under study. There was firstly the concept of a designed earth, particularly strong in the Judaeo-Christian tradition, in which man stood at the apex of creation, inhabiting a world expressly designed for his use. Before the advent of ecology, this was the West's attempt at an holistic conceptualization of the phenomena of the earth. Secondly, there can be recognized the theme of the influence of environment upon human culture, in particular where limits were apparently placed upon the development of human institutions and numbers, for (in consonance with the first concept) the whole earth should be fruitfully occupied by the human species. In this connection, Thomas Malthus stands out as a leading figure. Lastly, man as a modifier of nature was epitomized. In some cases the view was overtly optimistic, for man was seen as God's partner, putting the finishing touches to the divine work by draining swamps and reclaiming forests; in slightly later times George Perkins Marsh (1864) contradicted this view by pointing out the immense damage done in places with which he was familiar, especially around the Mediterranean Sea.

(*b*) *The nineteenth century and after.* A common element in all the ideas summarized above is their anthropocentricity (i.e. they are centred around man); and this notion becomes even more explicit in the nineteenth century, particularly as the idea of progress—implying especially control over nature—became manifest in the iron horses and stone palaces of industrialization. Hence forward, the West divorces itself most noticeably from the rest of the world: its major prophet of the time, Karl Marx, loudly asserts the superiority of man over nature, and states that, if only the proper institutions can be created, then environmental limitations of, for example, a Malthusian kind can be dismissed as evanescent shadows. But the very course of industrial growth removes men from any environment other than the constructed—cages for one species of great ape—and the reaction of existentialism and alienation is now powerful (Dubos, 1969). New-comb (1963) has shown of Becket's plays that the characters find their surroundings unfamiliar and hence hostile. The counter-reaction has come in the 'alternative society' with its emphasis on holism and the validity of instinctual behaviour. On a more conventional level, the concepts of ecology have placed in a new light the ideas of man's role in nature, his 'fight' against it and the assumption of unimpeded 'progress'. As Glacken (1970) remarks, man versus nature is an out-moded theme.

Today's perspectives on the relationship of man and environment partake strongly of these historic themes. The ecological view is, in its holism, heir to the idea of a designed earth, although it has moved far from the perfectibility of all man's activities. The study of econo-mics uses wealth as an index of progress, for nature exists to be trans-formed, if not into gold, then into Eurodollars. Ethologists must still argue how much behaviour is innate and how much learned, and hence must continue to debate the role of environment in moulding culture. And since throughout time some men have told others how they ought to behave, so ethics—environmental and otherwise—are still powerfully present.

2 *Ecological studies of man/environment relations*

Such an approach has not been confined to biologists; geographers, sociologists and doctors have all at times adopted what they imagine to be an ecological view. For our purposes, the contributions of biologists and geographers are of most interest and some of the ideas which they have in common are set out in section I. 2 (c).

(*a*) *The development of ecological views by biologists.* There is a long history of the study of plants and animals in their natural environment. Thus, either explicitly or implicitly, the influence of man has been discounted, by selecting unmodified habitats for study, by ignoring the effects of man or by being ignorant of them. In moving away from such a narrow view the discussion by Tansley (1949) of the various types of climax is important, for he specifically defined a vegetation type of deflected climax (plagioclimax) where the agent of diversion is man. The views he expressed stemmed from inspection of con-temporary vegetation, especially areas which were grazed or burned. His concept gained fresh impetus from the discoveries of Quaternary ecology, where the influence of man upon vegetation is shown through historic and pre-historic time. The seminal work of Iversen (1941) upon the Danish neolithic *landnam* phases has been followed with a host of studies from earlier periods (Dimbleby, 1962; Simmons,

1969; Smith, 1970) and later (Pennington, 1969; Oldfield, 1969) showing how agriculture and grazing have led to vegetation types which are essentially anthropogenetic.

Another critical development was the publication by Lindemann (1942) of his trophic-dynamic model of the ecosystem. Tansley's earlier coinage referred particularly to a spatial slice of biota to be inventoried; Lindemann's development referred to the functioning of the ecosystem with particular regard to the flow of energy, thus opening the way for the contemporary studies of production ecology that have been the main work of the International Biological Programme (Newbould 1964). In such a dynamic concept, man is but one of the components who may from time to time and place to place be the most important. The idea of ecological interactions as a system has led to the applications of systems analysis and computerization, both in conceptual development and practical application (Van Dyne, 1969; Watt, 1968). The role of population dynamics in characterizing ecosystems is of considerable interest (Elton, 1966; Lack, 1954; Solomon, 1969; Wynne-Edwards, 1962), especially where analogies with the human situation are drawn (Wynne-Edwards, 1965).

(*b*) *The contribution of geographers.* As an arbitrary but significant beginning, Humboldt may be chosen as a geographer-ecologist who frequently expressed opinions on the relations of man and his biotic environment. Likewise Marsh (1864), although not self-labelled a geographer, distinctly pointed to the ecological disintegration caused by certain cultures. In modern times, the role of the geographer seems to have been to adapt evolving biological concepts in order to enlarge and rectify the traditional man/environment school of geographical thought. In the steps of Barrows (1923), who elaborated his ideas to the point of calling geography 'human ecology', Eyre and Jones (1966) have tried to use modern studies to prove the essentially long-term viability of such a point of view. Eyre (1964) has also pointed out the fundamental contribution of ecology in setting limits to human activity. In the U.K. both Stoddart (1967) and Simmons (1966b; 1970) have used the conceptual basis of the ecosystem framework and its trophic-dynamic structure to point to its usefulness in modern geography; the work of Ackerman (1963) in the U.S.A. is along parallel lines.

(*c*) *Ecology, environment and conservation.* The essential basis of the ecological contribution is its holism, which led Fraser Darling (1963) to reject the subdivisions of plant, animal and human ecology in favour of a unitary 'ecology'. Following this the nature of natural systems is seen to be integrative, with all the components playing an essential role within the whole: energy flow, for example, is channelled so as to make solar energy pass through a large number of pathways and hence 'fuel' a diversity of organisms. Diversity produces ecological stability since population explosions are damped down by feedback mechanisms. By contrast, man's effects are to reduce the diversity of ecosystems, by simplifying them in order to crop them more easily, and hence bring about instability which can only be controlled by increasing the inputs of matter and energy made available by an industrial framework.

The implications for conservation are obvious. The planet is an ecosystem, and certain of the activities of one of its species threaten to disrupt the rest. Because of the integrated nature of the interaction between the living plants and their non-living surroundings, the tearing apart of the web of ecological relationships may bring about instabilities which even technological man cannot control.

3 *Economics and environment*

It is difficult to put a price on environment. This is why the main link between economics and environment has been through the value of the more tangible resources. Through its resources environment is affected principally in two ways: in the cropping or extraction of the resource, and in the use of biosphere (or atmosphere) as a sink for waste products. Traditionally, the model

$$resources \rightarrow man \rightarrow waste$$

has been viewed by economists as one in which both the resource pool and the garbage can are infinite. Additionally, the economic system erected on this material foundation has to grow continually in order to function properly. The role of price in adjusting supply and demand of resources is particularly important in capitalistic economies, but the basic exponentialist view is shared by Marxist economists also.

Of the various assumptions made by economists about the supply of resources (Lavering, 1968), none is more important than the idea of the technological 'fix'. Belief that improvements in technology and access to energy sources will constantly bring down costs (Barnett and Morse, 1963) enables the resource economists to foresee the use of even lower quality resources but in ever increasing amounts. In the case of minerals this is taken to its extreme by Brown (1954) who, postulating very cheap energy from nuclear sources, envisages ordinary granite as the source of all mineral resources. This attractive view of a cornucopian future for resources of the stock variety is challenged by Lavering (1969) and Cloud (1968; 1969) who dispute the assumptions about energy supply, technological advance and mineral distribution upon which it is founded.

In industrial societies continual growth of an economy means ever-expanding consumption and Galbraith (1967) suggests that the large corporations force consumption upon people, thus using resources and capital which might otherwise be used to improve the quality of both life and environment. Although rapidly expanding populations might be thought to be disadvantageous to many developing countries, Clark (1967) has put forward several arguments from the economic point of view which, he claims, show a rapidly rising level of population to be an asset to such a land. His arguments are, however, based upon the mistaken assumption that population growth affects no variables in the population resource/environment systems other than those it is supposed to influence favourably.

A major difference between ecological and economic views of the resource/environment complex is the time scale over which they operate. That of ecology tends to be very long even if not very precise, that of economics short—measured in periods of discount rates of capital and tied to the life-span of expensive equipment. That conflict should arise over management goals in the context of such different time spans is not surprising. Thus the most elaborate benefit/cost analysis, even if it can quantify all the costs, operates in a different milieu from the ecological assessment. Frequently, however, only the resource/man part of the model is accounted and the external diseconomies such as waste disposal and pollution, congestion and the like, are not priced (Mishan, 1967).

Ecologists and other 'environmentalist' people are therefore highly suspicious of the role of most of contemporary economics in regulating man's use of the environment. They see economists as totally transfixed by the attitudes of the nineteenth century and only the radical economics of for example Boulding (1966; 1970)—(see section III)—appear to be at all consonant with ecological thinking.

4 Ethology: the behavioural approach to environment attitudes

This approach looks at man/environment relations from within, as it attempts to see how both individuals and groups perceive their surroundings and hence, within the limits of their technology, how they act.

(*a*) *The individual*. A most interesting contribution is by Hall (1966), who discusses the role of culture as it affects each individual's cognition of space. He draws examples from European cultures of how, for example, Southern Europeans like to interact within breathing and touching distance of each other whereas in the north these actions would be considered impolite: witness the avoiding behaviour of people on a crowded street in Europe or North America. It is a far cry from this to environmental significance, but perhaps the tentative suggestions can be made that the kind of individual space which is desired may affect such matters as preference for housing and the tolerance of crowding. The Englishman's 'semi' and the Finn's wilderness cabin—which are quite significant in terms of environmental use and management—may thus be related to individual behaviour.

(*b*) *The group*. More significant than any individual, however, is the outcome of the perceptions of a group of people, particularly when acting through an institutional framework; and in these studies geographers have made some important contributions (see also Ch. 7). It is not clearly known what determines an individual's perception of his environment—although objective knowledge, fear, fantasy, imagination and prejudice are all involved—and still less that of the group (Wood, 1970). The complexity of the ecology of environment and of its social response means that the capacity of the human mind for formulating and solving complex problems (even where aided by the computer) is very small compared with the size of the problems whose solution is required for objectively rational behaviour in the real world. Naturally enough this state of bounded rationality leads to irrational behaviour: one instance is where a hazard is perceived but dismissed mentally in order to save the cost of adjusting to it (Burton, Kates and Snead, 1969; Saarinen, 1966).

Where different groups perceive an environment, such as one of the National Parks of England and Wales, in different ways the conflict over management aims and methods is inevitable. Some may see the resource in a utilitarian fashion as a yielder of lambs or potholes, others as a set of concrete objects to be planned or changed around; yet more have an emotional attachment to the landscape; and for a final group the value is largely symbolic, representing, perhaps, freedom from industrial drudgery. When all these groups are represented on a planning committee, for example, the scope for disagreement is wide.

Men see their environment, therefore, through the diverse lenses of their culture and behave accordingly. It might be as well to notice, following in the steps of Kant and Hume, that ecology and economics are also outgrowths of the mind, cultural frameworks imposed upon nature, so their claims to omniscience need to be viewed sceptically.

5 Ethics: normative behaviour towards environment

In the previous section, human behaviour towards its surroundings was considered as it is; ethics consist of some people telling others how they ought to behave, in this case environmentally. The question of values now enters the scene (Price, 1955), and it is notable that geographers have made very few contributions to the discussion because geography appears to have considered itself a value-free discipline.

(a) *Historical development of ethical attitudes.* In such a short discussion it is possible to isolate only a few main strands of thought. In the West, the Judaeo-Christian tradition is clearly of paramount importance, and in a much-quoted paper White (1967) has anathematized its environmental views. According to him, they are dominated by the command to humans to multiply and to subdue the earth. Thus has arisen the notion of nature as something to be conquered instead of lived alongside and thus the source of the world's current ecological ills. While asserting that Judaism and Christianity are unhealthily anthropocentric, writers like Dubos have challenged White's thesis, arguing that later developments stemming from industrialization, and gathered under the general heading of alienation, have been more crucial in contributing to what Nicholson (1968) has called environmental disease.

Investigations of the attitudes of the normative precepts of other cultures towards their environment are infrequent. These tend to be specific with regard to a particular environmental component such as animals in general (Buddhism) or cows in particular (Hinduism) and restricted in terms of environmental impact. The care of nature enshrined in Confucianism did not apparently prevent the ravaging of China in the attempt to grow enough food for its burgeoning population; the nature-cult of traditional Japan, based largely upon the beliefs of Shinto religion, seems to have been overlaid by Western values brought in with industrialization. A rich field, explored for the pre-nineteenth century West by Glacken (1966; 1967) and foreshadowed for some other places by Tuan (1967) exists for future literary-minded geographers.

(*b*) *Time present and time future*. White proposed that St. Francis should be the patron saint of ecologists; for practical purposes a more secular ethic must be put forward. Apart from a short set of lecture-sermons by Montefiore (1970), the most elaborate and useful discussion is by Black (1970) in which two Christianity-derived ideas are put in a form suited to wide acceptability. The key is the idea of stewardship: that man administers the use of the resources of the earth (and hence affects the systems of the whole planet) on behalf of someone else, not just for his own gain. For the religious, he is God's steward; for them also, and especially for the secular, he manages the earth for posterity. The concept of responsibility to descendants is very strong in many ethical traditions, and Black considers that it can be made the basis of an attitude entirely consonant with ecological limits—namely that the choices open to later generations are not reduced. This means especially the preservations of diversity of biota, habitats and cultures. Such a theme is echoed by Dasmann (1968a) who, while wishing to have his own Utopia, wants it to be different from that of other people's and would like to be free to visit theirs from time to time.

The single common feature of most ethical precepts now emerging as guidelines in environment-oriented value systems is that they are incompatible with the type of short-run economics that distributes resource and wealth in favour of the material gain of individuals.

II Conservation at regional and national scales

It cannot be said that there is generally a close and traceable relationship between the ideas outlined above and the practical management of resources that takes place under the label of 'conservation'. Nevertheless, much of this activity is ultimately motivated by one or more of the strands of thought discussed, however little this is realized. In practice two main abstractions are important at the scales now being discussed: that the world (or nature, or the environment) is beautiful, and that it is useful. Conflict between two such conceptions, both of which lay claim to be conservation-minded, is not difficult to find.

The question of scale is important in all discussion of resource use and management, for it determines the degree of integration between different uses of the environment. It is also important in deciding the institutional framework, whether this be a local voluntary society or the United Nations, and hence the perception of the habitat and its relation to larger wholes is likely to vary greatly.

1   *Conservation as preservation*

The value attached to old buildings, particular urban settings, historical sites, archaeological field monuments, wild animals and plants and their habitats, has led to strong movements, mostly in the West but increasingly elsewhere as well, for their protection. In this case protection means either their preservation in their present state or their restoration to some former degree of integrity. Numerous examples of such practices can be quoted. In Denmark, Newcomb (1967) shows how stress is laid on the preservation of visible monuments, either *in situ*, as with archaeological field monuments, or—if threatened—moved and rebuilt in an open-air museum setting. Farms with attendant fields and implements are to be seen at Hjerle Heide in Jylland, for example, and Den Gamle By in Aarhus is a famous set of urban relocations. The idea of the 'correct' buildings in the proper landscape combination of forests, heaths, lakes and farmland is strong in the Danish *Naturpark* planning concept where cultural landscapes are protected from unacceptable development.

This is paralleled in England and Wales by the institution of National Parks under the aegis of the National Parks and Access to the Countryside Act 1949. Valued landscapes, mostly uplands, are protected from uncontrolled developments of settlement, industry and caravan-sites. Agriculture and forestry, however, remain exempt from such control, although voluntary consultation often occurs. Regions of slightly lower value are designated as 'areas of outstanding natural beauty'.

In the case of wild life and interesting physiographic features, the Nature Conservancy of Great Britain administers a system of National Nature Reserves which vary in size and type from that of a large handkerchief to the whole of the Isle of Rhum. Controlled access and influence over the management of the land, when it is leased and not owned, aid the Conservancy's task of the perpetuation of particular habitats, flora and fauna. Their work is complemented by the numerous reserves owned or leased by Naturalists' Trusts, which are limited companies on a county or multi-county basis. At a rather larger scale, the idea of preservation motivated the passing of the Wilderness Act of 1964 in the U.S.A. This creates a National Wilderness Preservation System (Simmons, 1966a), initially of some 36,400 Km², but with additions to be approved by the Congress, in which protection of the existing environment from any forms of economic use (such as forestry, mining, water storage) is the paramount aim. The only permitted use is wilderness recreation, mostly in the form of back-country trips on foot or horseback. Most users want few facilities at all installed beyond perhaps an emergency telephone here and there. The natural environment may extend to lightning-set wildfires at intervals, especially in forest terrain, thus creating a problem for the appropriate Agency whether to try to control the fire or let it burn sundry campers as part of the 'wilderness experience'. As Lucas (1964) has shown, wilderness is mostly in the mind: an experience in which particular conventions play a key role.

Towns and cities are also subject to pressures for preservation. In the U.S.A. particular buildings may be preserved (as in Washington D.C. by the National Parks Service of the Federal Government) or reconstructions made of historically important structures which have vanished, as with several of the colonial settlements, including Williamsburg, Va., where the whole way of life is re-enacted; a similar venture is to be found at Upper Canada Village, near Morrisburg, Ontario. In Britain, the Civic Amenities Act 1967 made possible the declaration of historic parts of towns as Conservation Areas, to give special protection to the urban milieu.

The major difficulty with conservation, when thought of simply as preservation, lies in its essential desire to 'freeze' ecosystems and cultural landscapes at a particular point in time. Where a set of ecosystems is pristine and large enough to maintain itself, then there is justification for preventing change. In the case of cultural landscapes and man-altered ecosystems, the only reason for halting change is that one stage of development is more highly valued than another. Economic uses of the past have given us landscapes which are highly valued now—why should not today's economic uses be allowed to create their own landscapes? Thus runs the anti-preservationist argument (Simmons, 1967). Preservation has also suffered in the past from too strong a 'hands off' policy. Semi-natural ecosystems or ecosystems set in the matrix of a cultural landscape may well change away from the desired direction in the absence of management. Perpetuators of orchids in chalk grasslands must mow the grass to prevent it crowding out the rare flowers; the Luneberger Heide authorities subsidize the grazing of sheep to keep up the heathland. With buildings also, a use is necessary if they are to be properly maintained, and not every ancient building can be made into a museum. If public money is used for preservation then the public start to want to look inside!

In summary, preservation protects important parts of what Tunnard (1966) calls our 'cultural patrimony', but it is an insufficient answer to many of the current problems of man/environmental relationships.

2 *Conservation as wise use*

In both developed and undeveloped countries there exist schools of thought to whom unused resources represent a reproach. Nevertheless they reject the despoliation brought about by *laissez-faire* development with all its attendant waste. The idea of wise use therefore encompasses the development and use of resources for the greatest good of the greatest number, while minimizing unacceptable side-effects such as the loss of wild areas, despoliation of landscapes and the various types of pollution. The most usual application of this idea has been at a regional scale where, for example, resources have been utilized in order to create new prosperity in areas of economic decline, or to bring order and rationality into the resource use pattern of an area which was otherwise suffering from an excess of free enterprise. It is in projects of this nature that benefit/cost analysis finds one of its frequent uses, although the cost of insidious side-effects, such as the nutrient enrichment or eutrophication of water-bodies following agricultural intensification, is an example of an unaccounted result.

Development of resources on a regional basis may proceed with the management of one resource for a large area, in which all its uses are carefully evaluated and planned. The most common instance of such a resource is water, and studies such as those discussed in Ch. 3 serve as examples; particularly noteworthy are those of White (1963) on the Lower Mekong and Sewell (1965) on the Fraser River. Strategies of development in such cases emphasize the linkages between resource use in various places, and also their relation to the other resources of land, water and minerals that the region possesses or has available (see also Ch. 1). Such linkages are broadened out in a second type of regional development in which more than one resource is developed, particularly in order to alleviate the economic ills of its region. The T.V.A. scheme of 1933 (Clapp, 1955) is an obvious, and to some writers the only, example. Within the U.S.A. the Columbia Basin Project, dating from 1939, has also produced an economic revival based on an improved and diversified agriculture, services and recreational use (Macinko, 1963). Although power is often generated on such river-based schemes, it frequently does not benefit the region since it can easily be transferred outside it.

Regional schemes of a slightly different kind may be needed to cure more specific ills than general economic stagnation and decline. Livestock reduction on the Navajo range (Fonaroff, 1963) is an example which is proving difficult because of the cultural unacceptability of 'rational' stocking schemes; in Ontario the wake of Hurricane Hazel in 1954 brought a determination not only to manage certain watersheds as flood control areas against future eventualities of that order, but to combine such an enterprise with much-needed outdoor recreation land in multiple land-use schemes. A result has been the Ontario Conservation Authorities who successfully operate a number of multiple use schemes for water management, pollution control, outdoor recreation and nature conservation (Government of Ontario, n.d.; Pleva, 1961) (see also Chs. 1, 3 and 10).

In industrialized countries, regional plans for environmental quality are often made, and sometimes equated with conservation. Such plans are generally made in the face of urban sprawl or the relics of earlier, unplanned, industrial expansion. In Santa Clara County, California, the plan is to divide the urban area into cells with green wedges between and also to prevent too much takeover of agricultural land. Adequate open space for public open-air recreation is also a feature of this type of conservation (Santa Clara County, 1962). In the U.K. the Lea Valley scheme, in the London area, inaugurated by the Civic Trust in 1964, is to turn a series of derelict areas, playing fields, reservoirs and a canal into a linear park, thus transforming the quality of a whole environment (Civic Trust, 1964).

The idea of conservation as wise use has certain difficulties. At the outset it begs the question of what constitutes wisdom, and for whom. It may be especially difficult to arrive at a consensus of values which begird not only the present but future time also. More practical problems arise with the relation of population to the resources available to be developed: at this scale the planner usually has no say in the number of people for whom he is to plan—they are a given and exterior quality. The relationship of a region to larger units may be particularly problematical since the infrastructure of an institutional framework for achieving an integrated management of different resources is difficult to formulate (Caldwell, 1966). Even the Dutch experience with the *schap* method for predominantly recreational areas might not extend to a multi-resource project in a large region. In terms of the ecological criteria discussed earlier, however, wise use tends to be tied firmly to economic growth as the solver of all social problems, especially those of poverty; and the questioning of the virtues of the unlimited exponential 'growth' is an important outgrowth of the ecological tradition in conservation thought. In the underdeveloped countries development may well be necessary, but many regional development schemes in the West owe more to politicians seeking re-election than essential exploitation of resources.

3 Conservation as environmental management

This concept stems from the intense debates of the 1960s following the North American desire for programmes of management designed to produce 'environmental quality'. Environmental management attempts to integrate natural and social systems to the benefit of the latter and without detriment to the stability of the former. The uses of the environment have been defined (O'Riordan, 1971) to include those discussed as preservation and wise use, but they have a stronger ecological base which also forms a criterion of long-term acceptability, thus following Firey's (1960) precepts that resource processes should be ecologically possible, economically painful, and culturally acceptable. The environment is to be managed for the following purposes:

(i) protection of physical and mental health of the people
(ii) enhancement of economic gain
(iii) preservation of sensory and participatory pleasure.

The need for such an all-embracing structure for planning is clear, since many resource processes are so pervasive in their effects (Council of Europe, 1971). Only management of all aspects of environmental use on a national basis (or for big natural regions in a large country) is likely to have any long-term success. The strategies which stem from the concept of environmental management (which, it may be noted, is a highly anthropocentric idea: the earth is managed for man, while it is recognized that he is part of the interdependent community of the biosphere) involve conscious and rational choice from a variety of alternatives. While short-term economic objectives may be pursued, these should not preclude any future options and must be consonant with long-term ecological stability. In most cases this means avoiding environmental uses which reduce biological diversity. Invariably the ecological impact of development must be assessed, and it is a measure of the newness of the environmental management concept that few nations as yet have undertaken such a task. In Sweden there is now a National Environment Protection Board which performs such a function. The Federal government of the U.S.A. has passed legislation setting up an Environmental Protection Agency whose task it is to assess the impact of major industrial development, with special reference to pollution.

During the 1950s and 1960s it was realized that urban, industrial and agricultural wastes not only had reduced the quality of man's environment in many places but were also contributing large quantities of toxic substances (for example, lead and mercury) to man's environment and food (Rudd, 1963; Mellanby, 1967). Pollution of all kinds is, however, only a by-product of the use of the environment by man for resources, and attempts to clean up pollution will in many cases be unavailing unless considered in that context. All separate resource ecosystems such as agriculture, grazing, forestry and energy use (Dasmann, 1968b) will have to be *ecologically balanced*—i.e. stable in the long run—and additionally their integration on a national or multi-national basis must be achieved. The difficulties of formulating national management goals are immense; a lead, however, has been given by Odum (1969) in his 'strategy of ecosystem development'. He classifies ecosystems as manipulated by man into four categories:

(i)   urban-industrial land, where the built structures of man dominate the ecology

(ii)   intensively-used land, in particular that used for sedentary, highly productive agriculture as in Northern Europe or South-East Asia

(iii)   extensively-used land such as the great grasslands and savannas of the world, and also areas very little used such as the tundra and Boreal forest

(iv)   multiple use land, between (iii) and (iv), as in the forests of the U.S.A., which supply timber, water, recreation, grazing land and wildlife.

Odum suggests that for every management unit—interpreted in this essay as a nation—a proper balance of each of these units could be achieved, which would help to produce an ecologically stable use of land and water resources. So far, no work has emerged which sets out criteria for determining the proportion of each category (see also Ch. 1, section I (1)).

The aim of such work as that of environmental management, as envisaged by O'Riordan (1971), is to integrate the socio-economic systems of a country with their natural surroundings. The measurement of success is difficult: the quality of life for humans is not easy to quantify, since many would reject the assumption that it is merely a reflection of the Gross National Product (Boulding, 1970). Equally, ecological disintegration may be difficult to evaluate unless it is spectacular, as in the case of a total kill of biota by pollution, or as in the case of severe soil erosion. However, measures of diversity as set out by Margalef (1968) may prove to be useful. It is especially noteworthy that the approach outlined above is conscious of the key role of human population in the socio-economic systems and their impact upon the ecology of their environment. This applies both to economic resources such as food and minerals, and to non-economic qualities such as scenery and wildlife. More and more writers are stressing the notion that every nation has a carrying capacity for humans at a particular level of both material and environmental well-being.

III Conservation at the global scale

The activities of conservationists at the scales described above (section II) refer to the past and present. Global conservation is something for the future, and so the material of this section relates to ideas about time to come rather than actuality. The main difficulty about discussing man/environment/resource interactions at a global scale has been the great diversity of resource processes and the apparent impossibility of finding relevant abstractions to which the whole world could be related. However, Boulding (1966) has elaborated two models which serve that very purpose. The first he calls the 'cowboy' economy, and it characterizes the present situation in which resources are drawn from a seemingly infinite pool, used by man and converted into wastes which are disposed of into the biosphere and atmosphere. The whole system appears to depend upon continual growth for its health. The alternative is called the 'space-ship earth' model and derives from the idea of this fertile planet spinning alone in a presumably hostile universe. It is an isolated system and like a space ship its inhabitants are dependent for survival upon the life-support systems of their craft. Boulding desires a change to an equilibrium resource position in which the numbers of people are stabilized and all materials undergo recycling; hence wastes become raw materials and only irretrievable losses such as those of metals caused by friction are tolerated. Growth is no longer an overall aim of economic development, being replaced by stability as a desirable goal. This second model, at the global scale, is thus conformable to the closed system which, with the exception of energy flow, is a feature of our planet.

1   *The ecological basis of the space-ship earth concept*

The economic terms used by Boulding are of course isomorphic with the biological concept of a dynamic equilibrium in which the individual parts change but the overall level of activity (for example in terms of population levels or energy flow) remains stable at the carrying capacity of the habitat. It is not surprising, therefore, that the space-ship earth model has been enthusiastically adopted by ecologists as their vision of a long-term solution to environmental ills.

However sound its ecological base, it will require considerable adjustments in economic and ethical behaviour, some of which are discussed in sub-sections 2 and 3 below after an exposition of the major lineaments of the closed-system model.

(*a*) *A stable human population*. Since man is the user of materials and thus the creator of waste, his numbers are crucial. Such an analysis applies not only to metabolic resources such as food, but to the energy needed to maintain industrial societies, space for living and recreation, and intangibles such as the sight of wild birds (already fewer in number than those watching them at some popular spots) and the smell of unpolluted air. For these reasons and many others—as polemicized for example by Ehrlich (1968) and Ehrlich and Ehrlich (1970), and by many other neo-Malthusians (distinguished from the original and his immediate followers by their willingness to adopt medical methods of population control which the Reverend Thomas castigated as distinctly sinful)—the stabilization of the world's population and even its reduction becomes an urgent need. Indeed, it is the first priority. For different reasons a number of countries, especially the underdeveloped ones, also agree with this. Among the developed countries only the U.S.A. has really thought seriously about moving towards stability (Davoll, 1971; Day and Day, 1965); a British parliamentary report (Select Committee on Science and Technology, 1971) is ambiguous in its conclusions. In world terms, however, a retreat from the current 2 per cent per annum increase is seen as essential for the implementation of the space-ship earth type of economy.

(*b*) *Research on ecosystem stability*. Another need is for the identification and study of the ecological subdivisions of the planet and the effective preservation of such portions as were shown to be essential for the long-term stability of the planet's biological systems (Istock, 1969). Such systems are of course the life-support systems for the human inhabitants: we are completely dependent upon them.

(*c*) *Pollution control*. In order to stop poisoning parts of the biosphere and, from time to time, ourselves, the technology and institutional means to control all biologically significant wastes must be achieved. This includes all gaseous, liquid, solid and thermal wastes, and may be extended where culturally acceptable to include items such as noise and 'visual pollution'. The control and perhaps complete elimination of the emission of radioactive wastes is especially important because of the absence of any realistic biological thresholds (see also Ch. 8, section IV).

(*d*) *Cycling of materials.* This has obvious economic implications as well as ecological dimensions. Virtually all manufactured products would be re-used, lessening the demands upon non-renewable resources; all organic products would be mineralized and re-cycled. Management of renewable resources would be less intensive since exponential increases in production should not be required. Repercussions on economics would be considerable.

2   *Economic adjustments on the space-ship earth*

As indicated above, continued growth appears to be the only basis for the viability of contemporary economic systems. The adoption of the model under discussion would require the introduction of an equilibrium system, which would necessitate certain adjustments. Many of these cannot at present be foreseen, but some of the probabilities are considered below.

(*a*) *Consumption patterns.* It seems inevitable that the era of virtually unlimited increases in consumption of material resources in the developed countries will come to an end. Insufficient resources will make this certain, and the revolt against the consumer society now at an early stage may well work with this trend. Especially notable will be the trend away from disposability to long-lasting articles; items requiring rapid recycling will be relatively the most expensive because of the energy costs involved in the processes. Thus the Gross National Product will disappear as an ikon of stockbrokers and an indicator of social policy (Boulding, 1970). Furthermore, pricing will have to take account of the full social costs of the use of every article, including its re-processing costs.

Another social trend immediately visible in the developed countries will be the altered population structure, in which there will be a higher proportion of old people than in a rapidly expanding population. It is possible that this will cause difficulties in terms of shortage of productive labour of younger men, which has caused Japan to withdraw government support from family planning programmes, but in an era of diminished production and consumption this may well matter less. If more workers are needed then the value of the older person will increase, which will be a considerable social improvement. The continued and even enhanced dominance of older people in positions of authority may well, however, cause some social strain.

3   *The motivation for advocating the space-ship earth model*

The most obvious question is: what are the alternatives: can the present growth-oriented structure continue? The answer is a matter of belief rather than knowledge, but increasing numbers of writers tend to think that the alternative of the technological 'fix' is not to be relied upon

Beyond this, two major ethics present themselves. The first is the posterity argument outlined in section 1.5(b). This course is seen as the best way of leaving open the options to our successors: a mono-cultural earth managed only for feeding the largest number possible or producing for a ubiquitous ownership of two cars, and two tellies, together with unlimited Coca-Cola in no-deposit, no-return bottles, does not do so. Secondly, continually increased use of resources exacerbates the gap between rich and poor nations for it is in the former that most of the use occurs—in the U.S.A. in 1967 steel consumption was 634 kg per capita and in the same year in India it was 13 kg per capita (Davoll, 1971). A voluntary 'de-development' as envisaged by Ehrlich and Harriman (1971) could be one of the most effective ways of helping the third world to a quality of life scarcely possible by throwing them the crumbs from the table in the form of one per cent (or less) of GNP.

4   *Implementation*

The complications in implementing the space-ship earth model are immense. Even if agreement on desirability were reached, for example at the 1972 U.N. conference on the human environment in Stockholm, the practical details are of a mind-boggling order. Some steps are outlined in Ehrlich and Harriman (1971) but this scheme is probably too idealistic. One uncertainty that will slow any progress is the interaction between features of resource systems (for example, population, investment, pollution) once one element is changed drastically (for example, population growth). Recent work by Forrester (1971a; 1971b) has forecast that attainment of a world of equilibrium will require, starting in 1970, a reduction in world capital investment by 40 per cent reduction in birth rate of 50 per cent reduction in pollution rate of 50 per cent and reduction of food production of 20 per cent (Fig. 11.1). The accuracy of this analysis may be disputed but the figures presage considerable mental difficulties, especially for those in politics for whom a week is a long time. But it is worth repeating the question: what is the alternative?

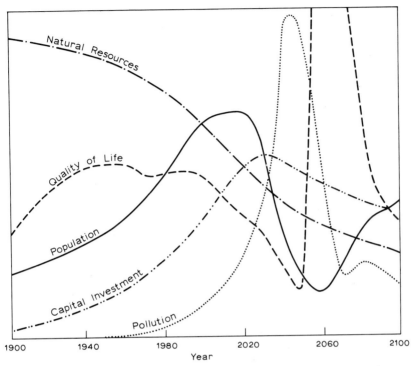

Fig. 11.1 An example of the 'world dynamic' interactions formulated for computer use by Forrester (1971a). In this sample, increased capital investment and decreasing resource use serve only to stave off a pollution crisis for a short period.

IV Interaction of geography and conservation

This essay has so far concentrated upon the various meanings, practical and theoretical, of the word 'conservation'. This final section will deal with generalizations affecting the relation of a few strands of geographical thought to the major types of conservation which have been discussed.

1 *Local and national scales*

Conservation at these scales is usually concerned with minimizing the impact of economic development but accepting the premises of economic growth. It is therefore for optimists who believe that a change of heart can bring about better environmental relationships

with the present socio-economic framework. Apart from their contributions as ordinary citizens (and therefore as active or apathetic
in about the same proportions no doubt), geographers may have
special professional contributions to make to conservation activity.
One of these lies particularly in the provision of information which can
produce a greater awareness of the importance of resources and environment. The significance of features of relict landscapes as symbolic of former man-environment relationships, for example, rather
than as mere monuments, may infuse a visitor with a sense of his past
as much as a total reconstruction (more so, if he is imaginative); the
study of land-use history likewise provides datum-lines against which
to measure the rapid changes enjoined by today's technology as well
as the causes and consequences of these alterations. Where man appears to have misused his environment, then geographers especially
should be aware of the multitude of factors which have probably
entered into the decision to undertake a particular resource process
and thus warn against the simplistic explanations sometimes offered
by proponents of instant ecology or omniscient Marxism.

Above all, as Clayton (1971) has stressed, the geographer is in a
position to see the interaction between physical systems and socio-
economic systems and to study the limits imposed by the latter. Each
physical system is capable of certain types of human manipulation
and has its flexibilities, but only people stoned on the heady smoke of
technology can believe that limits do not exist.

2   The global scale

Since world-order conservation is such a daunting proposition, only
the scared want to attempt it. It follows that those who advocate
space-ship earth economics are the pessimists who believe that present
trends will bring a drastic lowering of the quality of life at the very
least, and extinction of *Homo sapiens* not long afterwards. As ecologists
of the human species, geographers might be expected to contribute to
the understanding of the interaction of physical and socio-economic
systems on a world basis, so that the world's citizens and organizations
might take rational stock of their common resources. More likely perhaps is the geographer's potential to contribute an understanding of the
cultural factor in man/environment relationships. Ecologists are apt
to think of man as a rather homogeneous species, characterized by
the possession of technology and an exponential rate of increase;
geographers know the immense cultural variety over the earth, and
the differences in outlook and ideas which can enhance the diversity

provided by nature. Such a view can also be an antidote to those economists who persist in views that recall Wilde's definition of a cynic.

Whatever their views of the desirability of a dynamic equilibrium economic system, geographers must also play a part in the education of people in being a part of a single world. All resource processes, from food to pollution, are becoming trans-national and trans-continental, and solutions to problems, although lying eventually within the legislative ambit of nation states, must take heed of our neighbours and their needs and difficulties. Geographers will need no reading of Marshall McLuhan to remind them of the extent of their neighbourhood.

3 Values

Ultimately, this author believes that conservation is about human values—what a cultural group desires from its environment and what it will accept by way of change in order to get it. Whether values are expressed in emotional, economic, or ethical terms is not particularly relevant to the overall contention. Geography, by tradition, has been largely value-free in the sense that what happened was studied, badnesses were clucked over, and planners exhorted to do better next time. Normative values have therefore not been characteristic of our discipline, and the literature shows few instances of the role of cultural and ethical values in geography. (One should perhaps exempt geographers of the Berkeley school and Paul Wheatley.) At the time of writing (autumn 1971) there are signs of change. The emergence of radical geography in the U.S.A. (Smith, 1971) is heralded by Zelinsky's (1970) important paper in which he becomes the leading geographer to espouse publicly the space-ship concept and to chart the role for geography in the time 'beyond the exponentials'. His group are also examples of committed geographers who have thrown aside the traditional academic objectivity with its emphasis on another £10,000 grant from S.S.R.C. in order to find out what questions to ask.

The age of Aquarius, where Reich's (1970) Consciousness III puts much emphasis on new and non-exploitive attitudes towards our environment and resources, may be no passing fad. One of the major effects of conservation upon geographers (and the author would hazard a guess that it will be more definite than the effect of geographers upon conservation) may well be a move towards the type of publicly committed stance, based upon certain human values, that has so far been most noticeable among biologists. Who knows but the next decade or so may see a 'greening of geography', which will further erode the implicitly accepted view (discussed with tantalizing brevity by Pahl (1967)) that geography is a value-free discipline.

References

ACKERMANN, E. A. (1963): Where is a research frontier? *Annals of the Association of American Geographers* **53**, 429–40.

BARNETT, H. J. and C. MORSE (1963): *Scarcity and growth*. Baltimore.

BARROWS, H. H. (1923): Geography as human ecology. *Annals of the Association of American Geographers* **13**, 1–14.

BLACK, J. (1970): *The dominion of man*. Edinburgh.

BOULDING, K. E. (1966): The economics of the coming spaceship earth, in, H. JARRETT, editor: *Environmental quality in a growing economy*. Baltimore.

(1970): Fun and games with the gross national product—the role of misleading indicators in social policy, in H. W. HELFRICH, editor: *The environmental crisis*. Yale and London.

BROWN, H. (1954): *The challenge of man's future*. New York.

BURTON, I., R. KATES and R. SNEAD (1969): The human ecology of coastal flood hazard in megalopolis. *University of Chicago, Department of Geography, Research Paper* **115.**

CALDWELL, L. K. (1966): Administrative policies for environmental control, in F. FRASER DARLING and J. MILTON, editors: *Future environments of North America*. New York.

CIVIC TRUST (1964): *A Lea Valley regional park*. London.

CLAPP, G. R. (1955): *The TVA: an approach to the development of a region*. Chicago.

CLARK, C. (1967): *Population and land use*. London.

CLAYTON, K. M. (1971): Reality in conservation. *Geographical Magazine* **44,** 83–4.

CLOUD, P. (1968): Realities of mineral distribution. *Texas Quarterly* **11,** 103–26.

(1969): Mineral resources from the sea, in NATIONAL ACADEMY OF SCIENCES—NATIONAL RESEARCH COUNCIL. *Resources and Man*. San Francisco.

COUNCIL OF EUROPE (1971): *The management of the environment in tomorrow's Europe*. Strasbourg.

DASMANN, R. F. (1968a): *A different kind of country*. New York.

(1968b): *Environmental conservation* (second edition). New York.

DAY, L. and A. T. DAY (1965): *Too many Americans*. New York.

DAVOLL, J. (1971): Statement by the Conservation Society to the Minister of the Environment on the U.N. Stockholm 1972 Conference. London.

DIMBLEBY, G. W. (1962): The development of British heathlands and their soils. *Oxford Forestry Memoir* **23.**

DUBOS, R. (1969): *A theology of the earth*. Smithsonian Institute Lecture. Washington D.C.

EHRLICH, P. (1968): *The population bomb*. New York.
EHRLICH, P. and A. EHRLICH (1970): *Population, resources, environment: issues in human ecology*. San Francisco.
EHRLICH, P. and R. L. HARRIMAN (1971): *How to be a survivor*. New York.
ELTON, C. S. (1966): *The pattern of animal communities*. London.
EYRE, S. R. (1964): Determinism and the ecological approach to geography. *Geography* **49**, 369–76.
EYRE, S. R. and G. JONES (1966): *Geography as human ecology*. London.
FIREY, W. (1960): *Man, mind and land*. Glencoe, Ill.
FONAROFF, L. S. (1963): Conservation and stock reduction on the Navajo tribal range. *Geographical Review* **53**, 200–23.
FORRESTER, J. W. (1971a): *World Dynamics*. Cambridge, Mass.
  (1971b): Alternatives to catastrophe. Understanding the counterintuitive behaviour of social systems, part two. *The Ecologist* **1** (15), 16–23.
FRASER DARLING, F. (1963): The unity of ecology. *Advancement of Science* **20**, 297–306.
GALBRAITH, J. K. (1967): *The new industrial state*. New York.
GLACKEN, C. J. (1966): Reflections on the man-nature theme as a subject for study, *in* F. FRASER DARLING and J. MILTON, editors: *Future environments of North America*. New York.
  (1967): *Traces on the Rhodian shore*. Berkeley and Los Angeles.
  (1970): Man against nature: an outmoded concept, *in* H. W. HELFRICH, editor: *The environmental crisis*. Yale and London.
GOVERNMENT OF ONTARIO (n.d.): *Ontario's conservation authorities and the conservation authorities branch of the provincial government*. Toronto.
HALL, E. T. (1966): *The hidden dimension*. New York.
ISTOCK, C. (1969): A corollary to the dismal theorem. *Bioscience* **19**, 1079–81.
IVERSEN, J. (1941): Landnam i Danmarks stenalder. *Danmarks Geologiske Undersøgelser* **RII**, 66.
LACK, D. (1954): *The natural regulation of animal numbers*. Oxford.
LAVERING, D. (1968): Non-fuel mineral resources in the next century. *Texas Quarterly* **11**, 127–47.
  (1969): Mineral resources from the land, *in* NATIONAL ACADEMY OF SCIENCES—NATIONAL RESEARCH COUNCIL: *Resources and Man*. San Francisco.
LINDEMANN, R. (1942): The trophic-dynamic concept of ecology. *Ecology* **23**, 399–418.
LUCAS, R. C. (1964): Wilderness perception and use: the example of the Boundary Waters canoe area. *Natural Resources Journal* **3**, 394–411.
MACINKO, G. (1963): The Columbia basin project: expectations, realization, implications. *Geographical Review* **53**, 185–99.
MARGALEF, R. (1968): *Perspectives in ecological theory*. Chicago.
MARSH, G. P. (1864): *Man and nature; or physical geography as modified by human action*. Reprint edited by D. LOWENTHAL (1965). Harvard.
MELLANBY, K. (1967): *Pesticides and pollution*. London.
MISHAN, E. (1967): *The cost of economic growth*. London.
MONTEFIORE, H. (1970): *Can man survive?* London.
NEWBOULD, P. J. (1964): Production ecology and the International Biological Programme. *Geography* **49**, 98–104.

NEWCOMB, R. M. (1963): Becket's road. *Landscape* **13,** 29–30.
 (1967): Geographic aspects of the planned preservation of visible history in Denmark. *Annals of the Association of American Geographers* **57,** 462–80.
NICHOLSON, E. M. (1968): *The environmental revolution.* London.
ODUM, E. P. (1969): The strategy of ecosystem development. *Science* **164,** 262–70.
OLDFIELD, F. (1969): Pollen analysis and the history of land use. *Advancement of Science* **25,** 298–311.
O'RIORDAN, T. (1971): Environmental management, *in* C. BOARD et al., editors: *Progress in Geography* **3,** 171–231. London.
PAHL, R. E. (1967): Sociological models in geography, *in* R. CHORLEY and P. HAGGETT, editors: *Models in geography.* London.
PENNINGTON, W. (1969): *The history of British vegetation.* London.
PLEVA, E. G. (1961): Multiple purpose land and water districts in Ontario, *in* H. JARRETT, editor: *Comparisons in Resource Management.* Lincoln, Neb.
PRICE, E. T. (1955): Values and concepts in conservation. *Annals of the Association of American Geographers* **45,** 65–84.
PRINCE, H. C. (1971): Real, imaginary and abstract worlds of the past, *in* C. BOARD et al., editors: *Progress in Geography* **3,** 1–86. London.
REICH, C. (1970): *The greening of America.* New York.
RUDD, R. L. (1963): *Pesticides and the living landscape.* Madison.
SAARINEN, T. F. (1966): Perception of the drought hazard on the Great Plains. *University of Chicago, Department of Geography, Research Paper* **106.**
SANTA CLARA COUNTY (1962): *A plan for parks, recreation and open space.* San Jose.
SELECT COMMITTEE ON SCIENCE AND TECHNOLOGY (1971): *First Report: Session 1970–1971.* London.
SEWELL, W. R. D. (1965): Water management and floods in the Frazer River basin. *University of Chicago, Department of Geography, Research Paper,* **100.**
SIMMONS, I. G. (1966a): Wilderness in the mid-twentieth century U.S.A. *Town Planning Review* **36,** 249–56.
 (1966b): Ecology and land use. *Transactions of the Institute of British Geographers* **38,** 59–74.
 (1967): How do we plan for change? *Landscape* **17,** 16–18.
 (1969): Pollen diagrams from the North York Moors. *New Phytologist* **68,** 807–27.
 (1970): Land use ecology as a theme in biogeography. *Canadian Geographer* **14,** 309–22.
SMITH, A. G. (1970): The influence of Mesolithic and Neolithic man on British vegetation: a discussion, *in* D. WALKER and R. G. WEST, editors: *Studies in the vegetational history of the British Isles.* Cambridge.
SMITH, D. M. (1971): Radical geography—the next revolution? *Area* **3,** 153–7.
SOLOMON, M. E. (1969): Population dynamics. *Institute of Biology Studies in Biology* **18,** London.
STODDART, D. R. (1967): Organism and ecosystem as models in geography, *in* R. J. CHORLEY and P. HAGGETT, editors: *Models in geography.* London.
TANSLEY, A. G. (1949): *The British Islands and their vegetation.* Cambridge.

TUAN, YI-FU (1967): Attitudes towards environment: themes and approaches, *in* D. LOWENTHAL, editor: Environmental Perception and Behaviour. *University of Chicago, Department of Geography, Research Paper* 109.

TUNNARD, C. (1966): Preserving the cultural patrimony, *in* F. FRASER DARLING and J. MILTON, editors: *Future Environments of North America.* New York.

VAN DYNE, G., Editor (1969): *The ecosystem concept in natural resource management.* New York and London.

WATT, K. E. F. (1968): *Ecology and resource management.* New York.

WHITE, G. (1963): Contributions of geographical analysis to river basin development. *Geographical Journal* 129, 412–36.

WHITE, L. (1967): The historical roots of our ecological crisis. *Science* 155, 1203–7.

WOOD, L. J. (1970): Perception studies in geography. *Transactions of the Institute of British Geography* 50, 129–42.

WYNNE-EDWARDS, V. C. (1962): *Animal dispersion in relation to social behaviour.* Edinburgh.

(1965): Self-regulating systems in populations of animals. *Science* 147, 1543–8.

ZELINSKY, W. (1970): Beyond the exponentials: the role of geography in the great transition. *Economic Geography* 46, 498–535.

Bibliographic note

A number of books of reprinted articles have appeared which contain material germane to the theme of this essay. Relevant examples are:

BURTON, I. and R. W. KATES (1965): *Readings in resource management and conservation.* Chicago.

COX, G. R. (1969): *Readings in conservation ecology.* New York.

DAVIS, W. H. (1971): *Readings in human population ecology.* Englewood Cliffs.

DETWYLER, T. R. (1971): *Man's impact on environment.* New York.

HARDIN, G. (1968): *Population, evolution and birth control.* San Francisco.

SCIENTIFIC AMERICAN (1970): *The biosphere.* San Francisco.

(1971): *Man in the ecosphere.* San Francisco.

SMITH, R. L. (1972): *The ecology of man: an ecosystem approach.* New York.

Index